# Durability Design of Concrete Structures

# RILEM Reports

RILEM Reports are state-of-the-art reports prepared by international technical committees set up by RILEM, The International Union of Testing and Research Laboratories for Materials and Structures. More information about RILEM is given at the back of the book.

1 Soiling and Cleaning of Building Facades
  Report of Technical Committee 62-SCF. Edited by L. G. W. Verhoef
2 Corrosion of Steel in Concrete
  Report of Technical Committee 60-CSC. Edited by P. Schiessl
3 Fracture Mechanics of Concrete Structures: From Theory to Applications
  Report of Technical Committee 90-FMA. Edited by L. Elfgren
4 Geomembranes - Identification and Performance Testing
  Report of Technical Committee 103-MGH. Edited by A. Rollin and J. M. Rigo
5 Fracture Mechanics Test Methods for Concrete
  Report of Technical Committee 89-FMT. Edited by S. P. Shah and A. Carpinteri
6 Recycling of Demolished Concrete and Masonry
  Report of Technical Committee 37-DRC. Edited by T. C. Hansen
7 Fly Ash in Concrete - Properties and Performance
  Report of Technical Committee 67-FAB. Edited by K. Wesche
8 Creep in Timber Structures
  Report of TC 112-TSC. Edited by P. Morlier.
9 Disaster Planning, Structural Assessment, Demolition and Recycling
  Report of TC 121-DRG. Edited by C. De Pauw and E. K. Lauritzen
10 Applications of Admixtures in Concrete
  Report of TC 84-AAC. Edited by A. M. Paillere
11 Interfacial Transition Zone in Concrete
  Report of TC 108-ICC. Edited by J.-C. Maso
12 Performance Criteria for Concrete Durability
  Report of TC 116-PCD. Edited by H. K. Hilsdorf and J. Kropp
13 Ice and Construction
  Report of TC 118-IC. Edited by L. Makkonen
14 Durability Design of Concrete Structures
  Report of TC 130-CSL. Edited by A. Sarja and E. Vesikari

# Durability Design of Concrete Structures

## Report of RILEM Technical Committee 130-CSL

RILEM
(The International Union of Testing and Research Laboratories for Materials and Structures)

Edited by

## A. Sarja and E. Vesikari

Technical Research Centre of Finland,
VTT Building Technology, Espoo, Finland

CRC Press
Taylor & Francis Group
Boca Raton London New York

CRC Press is an imprint of the
Taylor & Francis Group, an **informa** business
A TAYLOR & FRANCIS BOOK

CRC Press
Taylor & Francis Group
6000 Broken Sound Parkway NW, Suite 300
Boca Raton, FL 33487-2742

First issued in paperback 2019

© 1996 RILEM
CRC Press is an imprint of Taylor & Francis Group, an Informa business

No claim to original U.S. Government works

ISBN-13: 978-0-419-21410-6 (hbk)
ISBN-13: 978-0-367-86537-5 (pbk)

**Visit the Taylor & Francis Web site at**
**http://www.taylorandfrancis.com**

**and the CRC Press Web site at**
**http://www.crcpress.com**

Typeset in 11/13 Palatino by Words & Graphics Ltd., Anstey, Leicester

A catalogue record for this book is available from the British Library

Library of Congress Catalog Card Number: 95–71380

**Publisher's Note**
The publisher has gone to great lengths to ensure the quality of this reprint
but points out that some imperfections in the original may be apparent

# Contents

# Preface

Control of the durability of concrete structures has gained increasing importance at the design and maintenance stages of these structures. Most of the current durability problems of concrete structures in, for example, bridges, dams and façades of buildings could have been avoided with systematic durability design. Such design, however, requires both an overall methodology and detailed calculation models of actual deterioration processes. Safety requirements and margins must be defined for service life. This report represents the first systematic attempt to introduce into structural design both a general theory of structural reliability and existing calculation models for the commonest degradation processes.

In fact, the introduction of systematic durability design will mean better utilization of the existing research results for systematized control of the service life of concrete structures during design. These pioneering attempts at systematic durability design will help to formulate new ideas and identify the needs for further research of degradation processes and their calculation models. Durability calculations enable not only priority ranking of materials and structural factors, but also produce numerical values of factors for the intended service life.

The idea of this report is to present the methodology of durability design in such a way that it can be combined with traditional mechanical design. This procedure utilizes current results with regard to safety and performance under static, fatigue and impact loading. The greatest change is the highly improved design regarding durability. Through durability design, all durability parameters such as the concrete cover, durability properties of materials, amount of reinforcement and dimensions of the structures are calculated, taking into account in each case actual degradation processes and, where necessary, their interaction.

In order to understand the design methodology and calculation methods, a fairly extensive introduction is given to the statistical background. The reliability theory of structures is applied to durability safety principles. A number of new definitions are also presented.

The planning and writing of this report were done in close cooperation by the RILEM TC 130-CSL working team comprising the following persons:

Prof. Asko Sarja (Chairman)
Dr Carmen Andrade
Mr A. J. M. Siemes
Mr Erkki Vesikari (Technical Secretary)

The members of RILEM TC 130-CSL were:

Prof. Asko Sarja, Finland (Chairman)
Mr A. J. M. Siemes, The Netherlands
Prof. Lars Sentler, Sweden
Prof. F. H. Wittman, Switzerland
Prof. Eberhard Berndt, Germany
Prof. Erik Sellevold, Norway
Prof. Peter Schießl, Germany
Dr Carmen Andrade, Spain
Prof. H. K. Hilsdorf, Germany
Director Robert Serene, Switzerland
Prof. Fuminori Tomosawa, Japan
Mr Erkki Vesikari, Finland (Technical Secretary)

Important ideas and comments were contributed by several members of RILEM TC 130-CSL and related Technical Committees in RILEM and CEB, especially Dr Christer Sjöström, supervisor of the working commission, Prof. H. K. Hilsdorf, Dr C. D. Pomeroy and Prof. Peter Schießl. I should like to express my gratitude to them, to all the members of TC 130-CSL and to all those who contributed to our work.

*Asko Sarja*
*Chairman of RILEM TC 130-CSL*
*Espoo, August 1994*

# 1

# Introduction

## 1.1 BACKGROUND

Traditionally the durability design of concrete structures is based on implicit rules for materials, material compositions, working conditions, structural dimensions, etc. Examples of such 'deem-to-satisfy' rules are the requirements for minimum concrete cover, maximum water/cement ratio, minimum cement content, crack limitation, air content, cement type and coatings on concrete. These rules are sometimes related to the type of environmental exposure such as indoor climate, wet exposure, presence of frost and deicing salts, sea water and so on. The purpose of all these rules has been to secure robustness for structures, although no clear definition for service life has been presented.

Modern building codes will increasingly be based on the performance of buildings. It must be ensured that this performance exists throughout the service life of the building. With deem-to-satisfy rules it is not possible to give an explicit relationship between performance and service life. For concrete, but also for other building materials, these relationships are not yet available as design tools. They have to be developed.

Especially in the 1990s, clients and owners of buildings have shown increasing interest in setting requirements for the service life of structures. This has been a natural consequence of a greater awareness of quality and costs of buildings. Under-standing durability as an essential part of the quality of a building has become self-evident. Also increasingly emphasized is the fact that total costs comprise not only immediate construction costs, but also costs of maintenance and repair. As a result, durability and service life aspects have been stressed in contract briefs. New methods for more accurate durability design of structures have been demanded.

On the other hand, the vast research of the 1970s to the 1990s on concrete durability has produced reliable information on deterioration processes. Based on this knowledge it is now possible to incorporate durability even in the mechanical design of concrete structures.

A structural designer must be able to prove the fulfilment of a service life requirement. For him a simple durability model showing the performance over time or the service life as a function of appropriate design parameters is a valuable tool. With the aid of durability models a designer can make decisions on the required dimensions and material specifications for structures with a service life requirement. Even relatively rough durability models will give a rational basis for this kind of durability design.

Many environmental factors affect the degradation mechanisms of materials, but their precise influence is difficult to predict as they vary greatly locally. Also the properties of the building materials themselves may vary substantially. Because of these and other factors which cause increased scatter, the performance and service life of a structure should preferably be treated stochastically. This means that not only average values, but also distributions, are considered. Stochastic treatment of design problems takes into account the real nature of structural performance, making a reliable design of structures possible.

## 1.2 AIM OF THE WORK

The aim of the RILEM commission TC 130–CSL has been to work out the theoretical background and the design procedure of the durability design of concrete structures and to present a selection of durability models which are suited for such a design. Additionally, the aim was to present some examples of practical durability design. The report is aimed at assisting designers at the design stage of concrete structures for both normal outdoor conditions and aggressive environments. The purpose is to show how the results of materials research can be transferred to the design of structures.

To this end the work of TC 130–CSL included the following tasks:

1. examination of methods for incorporating degradation of materials into the structural design of structures;
2. study and selection of durability models for the commonest degradation factors of concrete structures;
3. study of stochastic methods of durability design;
4. laying down design procedures for concrete structures with elucidating examples.

It is hoped that this report will also serve as a link towards the development of new design codes of concrete structures. It is evident that this first attempt cannot be complete with all relevant degradations, environmental conditions and types of concrete structures, but by exposing the needs it will pave the way for further durability research and development of design.

The work of RILEM TC 130 has been linked to that of CEB Commission V, Working Group 1 'Calculation models' and Working Group 2 'Environments'. The role of the CEB work is more formal and directed towards new model codes. The character of the RILEM work can be described as freer, less formal, demonstrative and prenormative.

# 2

# Definitions and explanation of durability concepts

## 2.1 DEFINITIONS

| | |
|---|---|
| Degradation | Gradual decrease in performance of a material or a structure. |
| Degradation factor | Any of the group of external factors, including weathering, biological, stress, incompatibility and use, that adversely affect the performance of building materials and components (Masters and Brandt, 1989). |
| Degradation mechanism | The sequence of chemical, mechanical or physical changes that lead to detrimental changes in one or more properties of building materials or components when exposed to one or a combination of degradation factors (Masters and Brandt, 1989). |
| Degradation model | Mathematical function showing degradation with time (or age). |
| Degradation process | Time-dependent mechanism leading to degradation. |
| Design service life | Service life used in the design of structures to take into account the time-related scatter, and to provide a required safety against falling below the target service life. |

| | |
|---|---|
| Deterioration | The process of becoming impaired in quality or value. |
| Durability | The capability of a building, assembly, component, structure or product to maintain minimum performance over at least a specified time under the influence of degradation factors. |
| Durability design parameter | A property of a material, a dimension of a structure or its detail or the location of reinforcement influencing the durability of a structure. |
| Durability model | Degradation, performance or service life model of calculation. |
| Durability failure | Exceeding the maximum degradation or falling below the minimum performance. |
| Failure | Exceeding or falling below the limit state (serviceability limit or ultimate limit). |
| Failure probability | The probability of failure occurring. |
| Lifetime | Period of time from installation to the moment of examination. |
| Lifetime safety factor | Coefficient by which the target service life is multiplied to obtain the corresponding design service life. |
| Limit state of durability | Minimum acceptable state of performance or maximum acceptable state of degradation. The limit state may be set with regard to the ultimate limit or the serviceability limit. |
| Mechanical design parameter | A property of a material, a dimension of a structure or its detail, or the location of reinforcement influencing the mechanical resistance of a structure. |
| Performance | Measure to which the structure fulfils a certain function. |
| Performance model | Mathematical function showing performance with time (or age). |
| Risk | Multiplication of the probability of failure by the amount of damage. |

| | |
|---|---|
| Service life | Period of time after manufacturing or installation during which all essential properties meet or exceed minimum acceptable values, when routinely maintained (Masters and Brandt, 1989). |
| Service life model | Mathematical function for evaluating the service life. |
| Serviceability | Capacity of a building, assembly, component, product or structure to perform the service function(s) for which it is designed and used. |
| Serviceability limit state | State which corresponds to conditions beyond which specified service requirements for a structure or structural element are no longer met (CEN, 1994). |
| Target service life | Required service life imposed by general rules, the client or the owner of the structure. |
| Ultimate limit state | State associated with collapse, or other similar forms of structural failure (CEN, 1994). |

## 2.2 EXPLANATION OF SOME DURABILITY CONCEPTS

### 2.2.1 Performance and degradation

Performance is generally understood as behaviour related to use. In principle the performance can be related to bearing capacity, stability, safety in use, tightness, hygrothermal properties, acoustic properties, visual appearance etc. (RILEM, 1986). In the context of this report, we assume that performance is a quantifiable property.

Performance is always a function of time, which is why expressions like 'over time' or 'with time' are commonly added to the concept of performance. When time is considered in the evaluation of performance various external factors, called degradation factors, take on great significance. In this way performance is linked to the concept of durability. Durability is the property expressing the ability to maintain the required performance.

Degradation is by definition the gradual decrease in performance over time. Often degradation can be understood as the opposite, or inverse, of performance. Degradation serves an optional way to treat performance problems.

The concepts of performance and degradation over time can be applied at different levels: (1) buildings, (2) structural components, and (3) materials. It is important to identify the level used. However, there may be interactions between levels. If, for instance, the load-bearing capacity of a concrete column or a beam is studied, the problem is actually treated at the structural component level. However, in the long term the load-bearing capacity will be dependent on the degradation of concrete and steel. Thus the performance over time at structural component level must be evaluated by first analysing the rate of change in performance at material level.

The minimum acceptable values for performance (or maximum acceptable values for degradation) are called durability limit states. The limit state is a performance requirement critical to the service life, which can be set with regard to either the ultimate limit or the serviceability limit.

### 2.2.2 Service life

Service life is the period of time after manufacture during which the performance requirements are fulfilled. As with performance, service life can be treated at different levels. The necessary actions taken at the end of service life depend on the level applied. At building level the end of service life would normally entail complete renovation, reconstruction or rejection of the building. At structural component or material level it would mean replacement or major repair of those components or materials.

On the other hand, the problem of service life can be approached from at least three different aspects:

1. technical
2. functional
3. economic.

Depending on the aspect, one talks about technical, functional or economic service life. Different aspects give rise to different requirements for the object at hand.

Technical requirements are demands for technical performance. Depending on the level of treatment, they include requirements for the structural integrity of buildings, the load-bearing capacity of structures, and/or the strength of materials. Most of these requirements are set in the codes and standards.

Functional requirements are set in relation to the normal use of buildings or structures. For instance, the width and height of a bridge must meet the requirements of traffic both on and under the bridge. In this case the length of service life is not primarily dependent on the technical condition of the structure, but rather on its development in traffic. From the economic point of view a building, structural component or material is treated as an investment and requirements are set on the basis of profitability.

In the context of this report the aspect of service life problems is mainly technical. The technical point of view covers the following sub-aspects:

1. mechanical and other structural performance
2. serviceability and convenience in use
3. aesthetics.

The primary focus is on the mechanical and other structural functioning of concrete structures. The load-bearing capacity of structures can be violated by the degradation of concrete and reinforcement. Structures must be designed so that the minimum safety level is secured during the intended service life despite degradation and ageing of materials.

Defects in materials may also lead to poor serviceability or inconvenience in the use of a structure. For instance, disintegration of concrete in a pavement may cause inconvenient vibration and impacts in vehicles.

Aesthetic aspects are included in the technical requirements if the aesthetic defects of structures are due to degradation or ageing of materials. Then the question of aesthetics can be treated in technical terms.

The exact definition of service life is obscured by the fact that maintenance routines are performed during the service life of a structure. Maintenance can influence the length of service life and should therefore be included in the definition. Thus the addition 'when routinely maintained' as stated in the definition of TC 71-PSL is recommended (Masters and Brandt, 1989).

In the design of structures, the requirement of service life imposed by the client or owner of the building is called the target service life.

### 2.2.3 Probability of failure

In stochastic durability design, not only target service life but also the definition of maximum allowable probability of not reaching the target service life is necessary. It is called the probability of failure. On the other hand, probability of failure can be defined as the probability of exceeding or falling below a certain limit state, which may be an ultimate limit state or a serviceability limit state.

When failure is caused by degradation of materials, the term 'durability failure' is used as distinct from 'mechanical failure', which is caused by mecahnical loads. However, durability failure may be a partial reason for mechanical failure.

The required probability of failure depends on how the event of failure is defined, and what the consequences of such a failure would be. If the failure is expected to bring serious consequences, the maximum allowable failure probability must, of course, be small. For evaluating the consequences of failure the concept of risk may be valuable. The risk is defined as the multiplication of the probability of failure by the amount of damage (Kraker, de Tichler and Vrouwenvelder, 1982).

In general, when evaluating the required probability of failure, social, economic and environmental criteria should be considered. Social criteria should include:

1. the social importance of a building or structure;

2. the consequences of failure (the number of human lives at stake, etc.);
3. difficulty in the evaluation of the risk level.

Economic criteria may be important. Any interruption in the production of a factory due to a failure may cause considerable economic losses in comparison to the construction cost of the structures.

Environmental and ecological criteria must also be considered. In some cases, structural damage may lead to a serious environmental accident. In normal buildings a long and secure service life is also best in line with ecological principles whereby the consumption of building materials is small and the emissions and energy consumption during the manufacture of materials are minimized.

In the design phase the uncertainties with respect to the final quality of the building are greater than for existing buildings. This implies that for existing buildings the safety margins can be smaller than for buildings in the design phase. For both, the probability of failure is in principle the same. For the same reason it can be stated that, in assessing the effect of degradations on structural behaviour, the same theory can be applied to existing structures as to structures in the design phase. The only difference will be that in the end, the safety margins for existing structures will be lower than for structures in the design phase.

# 3

# Methods of durability design

## 3.1 GENERAL

The theory of durability design is in principle based on the theory of safety (or structural reliability) traditionally used in structural design. In this context safety denotes the capacity of a structure to resist, with a sufficient degree of certainty, the occurrence of failure in consequence of various potential hazards to which the structure is exposed. The theory, however, has hitherto been confined more particularly to problems in which time plays only a subordinate part. Now the use of this technique is increasingly advocated for dealing also with durability and service life problems (Siemes, Vrouwenvelder and van den Beukel, 1985).

Although traditionally the methodology of safety has been almost exclusively applied to studies of structural mechanics, the method is by no means restricted to such design problems. The methods presented in the following can in principle be applied to any performance problems dealing with structures or materials.

A new feature in the theory of safety is the incorporation of time into the design problems. It allows the possibility of treating degradation of materials as an essential part of the problem. Safety against failure (falling below the performance requirement) is a function of time. Designing a structure with the required safety now includes a requirement of time during which the safety requirement must be fulfilled. In other words a requirement for the service life must be imposed.

In the design of structures the required service life is called the target service life. The level of safety is expressed as the maximum allowable failure probability.


## 3.2 THEORY OF FAILURE PROBABILITY AND SERVICE LIFE

The simplest mathematical model for describing the event 'failure' comprises a load variable $S$ and a resistance variable $R$. In principle the variables $S$ and $R$ can be any quantities and expressed in any units. The only requirement is that they are commensurable.

If $R$ and $S$ are independent of time, the event 'failure' can be expressed as follows: (Kraker, de Tichler and Vrouwenvelder, 1982):

$$\{\text{failure}\} = \{R < S\} \tag{3.1}$$

In other words, the failure occurs if the resistance is smaller than the load.

The failure probability $P_f$ is now defined as the probability of that 'failure':

$$P_f = P\{R < S\} \tag{3.2}$$

Either the resistance $R$ or the load $S$ or both can be time-dependent quantities. Thus the failure probability is also a time-dependent quantity. Considering $R(\tau)$ and $S(\tau)$ are instantaneous physical values of the resistance and the load at the moment $\tau$ the failure probability in a lifetime $t$ could be defined as

$$P_f(t) = P\{R(\tau) < S(\tau)\} \quad \text{for all } \tau \leq t \tag{3.3a}$$

The determination of the function $P_f(t)$ according to equation (3.3a) is mathematically difficult. Usually the resistance and the load cannot be treated as instantaneous physical values. That is why $R$ and $S$ are considered to be stochastic quantities with time-dependent or constant density distributions. By this means the failure probability can usually be defined as

$$P_f(t) = P\{R(t) < S(t)\} \tag{3.3b}$$

According to definition (3.3b) the failure probability increases continuously with time as schematically presented in Figure 3.1. At the moment $t = 0$ the density distributions of the load and the resistance are far apart and the failure probability is small at first. With time the distributions approach each other, forming an overlapping area of increasing size. The overlapping area

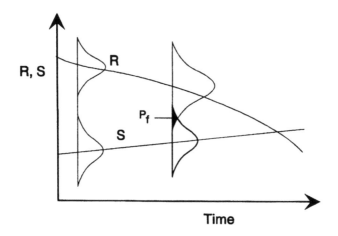

**Figure 3.1** The increase of failure probability. Illustrative presentation.

illustrates the failure probability.

The function $P_f(t)$ has the character of a distribution function. If the service life is defined so that the event '$t_L < t$' is identical with the event 'failure in lifetime $t$' the distribution function of service life can be defined as

$$F_L(t) = P\{t_L < t\} = P_f(t) \tag{3.4}$$

where $F_L$ is the cumulative distribution of service life.

The probability density function can be determined as the derivative of the distribution function:

$$f_L(t) = \frac{d}{dt}F_L(t)$$

(3.5)

At a certain moment of time the probability of failure can be determined as the sum of products of two probabilities: (1) the probability that $R < S$, at $S = s$, and (2) the probability that $S = s$, extended for the whole range of $S$:

$$P_f = \sum_{\forall_s} P\{R < S|S = s\}P\{S = s\}$$

(3.6a)

Considering continuous distributions the failure probability $P_f$ at a certain moment of time can be determined using the convolution integral:

$$P_f = \int_{-\infty}^{\infty} F_R(s)f_S(s)ds$$

(3.6b)

where $F_R(s) =$ the distribution function of $R$,
$f_S(s) =$ the probability density function of $S$, and
$s =$ the common quantity or measure of $R$ and $S$.

The general solution of the convolution integral with time-dependent distributions of $R$ and $S$ may be troublesome. The straightforward solution of the integral is only available in a few cases, i.e. when the distributions of $R$ and $S$ are normal. However, the integral can be solved by approximative numerical methods. The distribution of service life can be obtained by calculating the failure probability values at different moments of time, e.g. $t = 0, 10, 20, 30$, etc. years.

## 3.3 BASIC FORMULATION OF DURABILITY DESIGN

### 3.3.1 Principles and methods of design

The basic formulae of durability design can be written according to two optional principles:

1. performance principle
2. service life principle.

When the performance principle is used, the basic design formula is written by setting the load $S$ into a relationship with the performance $R$. The performance, evaluated by a performance model, must be greater than the required load. The performance can be related to the mechanics, physics, chemistry, and functions of the use, or to aesthetics. The load can be either mechanical or environmental.

When the service life principle is used, the service life $t_L$, evaluated by a service life model, must be greater than the required target service life $t_g$.

Normally both principles lead to the same results. Which principle is selected depends much on how the design problem is set up, what kind of durability models are available (cf. Chapter 7) and what is the easiest way to treat the problem mathematically.

The method of durability design can be (1) deterministic, (2) stochastic or (3) lifetime safety factor. The principles of these methods are explained below.

### 3.3.2 Deterministic design method

In deterministic durability design, the load, resistance and service life are used as deterministic quantities. The distributions of these functions are not considered. According to the performance principle the design formula is written as

$$R(t_g) - S(t_g) > 0 \qquad (3.7)$$

where $t_g$ is the target service life.

The load $S$ or resistance $R$, or both, are time-related functions. Degradation and performance models (cf. Chapters 7 and 8) are used as these functions. Design parameters such as structural dimensions and material specifications and environmental coefficients are incorporated into these functions.

The corresponding design formula is written according to the service life principle as follows:

$$t_L - t_g > 0 \qquad\qquad (3.8)$$

where $t_L$ is the service life function. Service life models including design parameters (cf. Chapters 7 and 8) are used as this function.

The design of structures is performed by selecting an appropriate combination of values for design parameters in such a way that the conditions of equations (3.7) and (3.8) are fulfilled.

When using the service life principle it is important to note that the performance requirement is incorporated into the formula for service life. This is why the performance concept and service life concept lead to a similar design formulation, i.e. the results of the design should be the same irrespective of the method used.


### 3.3.3 Stochastic design method

In stochastic durability design, the distributions of load, response and service life are also taken into account. The condition is expressed as the probability that the design formula is not true. The design formula can be written according to the performance principle or the service life principle in basically the same way as in deterministic design. However, a requirement for the maximum allowable failure probability is added to the final condition.

According to the performance principle the following requirement must be fulfilled: the probability of the resistance of the structure being smaller than the load within the service period is smaller than a certain allowable failure probability.

Mathematically the requirement is expressed as

$$P\{\text{failure}\}_{t_g} = P\{R - S < 0\}_{t_g} < P_{\text{fmax}} \qquad (3.9)$$

where $P\{\text{failure}\}_{t_g}$ = the probability of failure of the structure within $t_g$, and

$P_{\text{fmax}}$ = the maximum allowable failure probability.

The problem can be solved when the distributions of the load and the resistance are known. In section 4.2 a solution is presented for cases where the load and resistance are normally distributed.

When the service life principle is applied the requirement is set as follows: the probability that the service life of a structure is shorter than the target life is smaller than a certain allowable failure probability. Mathematically the condition is written as

$$P\{\text{failure}\}_{t_g} = P\{t_L < t_g\} < P_{\text{fmax}} \qquad (3.10)$$

The problem can be solved if the distribution of service life is known. If the form of distribution is not known, it must be assumed to follow some known distribution type. A solution for cases where the distribution of service life is assumed to be log-normal is presented in section 4.3.

## 3.3.4 Lifetime safety factor method

When the formulae for load, resistance and service life are complex and many degradation factors affect the performance of structures, application of the above stochastic design method may be difficult. In such cases it may be reasonable to apply the lifetime safety factor method. Although the method is based on the theory of safety and reliability, formulation of the design procedure returns to the deterministic form. This is possible by changing the requirement of target service life to the corresponding requirement of design service life. In practice the design service life is determined by multiplying the target service life by a lifetime safety factor:

$$t_d = \gamma_t t_g \tag{3.11}$$

where $t_d$ = the design service life,
$\gamma_t$ = the lifetime safety factor and
$t_g$ = target service life.

The design formulae can then be written by applying either the performance principle or the service life principle as follows:

$$R(t_d) - S(t_d) \geq 0 \tag{3.12}$$

$$t_L - t_d > 0 \tag{3.13}$$

The value of the lifetime safety factor depends on the maximum allowable failure probability. The lifetime safety factor must be calibrated with the results of stochastic design methods as presented in Chapter 5.

The lifetime safety factor method is used in structural durability design as discussed in Chapter 6. The application of the method is especially justified in structural design because the stochasticity of loads and strengths is also taken into account by safety factors, the method in this way being in principle familiar to mechanical structural design.

# 4

# Examples of durability design by stochastic methods

## 4.1 GENERAL

In order to use stochastic design methods, some assumptions must be made concerning the form of distributions. Distribution types that can be used for the evaluation of service life or performance of structures include:

1. normal (Gaussian) distribution
2. log-normal distribution
3. exponential distribution
4. Weibull distribution
5. gamma distribution.

When the performance principle is applied, the commonest assumption is that either the load or the resistance, or both, are normally distributed. When the service life principle is used, the distribution of service life is often assumed to be log-normal, i.e. normal on a logarithmic time scale.

## 4.2 DESIGN WITH PERFORMANCE PRINCIPLE WHEN $R$ AND $S$ ARE NORMALLY DISTRIBUTED

### 4.2.1 Theory

When the performance principle is used in durability design and the resistance, R, and the load, S, are normally distributed quantities, the failure probability can be determined using the test index $\beta$:

$$\beta(t) = \frac{\mu[R,t] - \mu[S,t]}{(\sigma^2[R,t] - \sigma^2[S,t])^{1/2}} \tag{4.1}$$

where $\mu$ denotes the mean and $\sigma$ the standard deviation.

The test index $\beta$ is (0,1)-normally distributed. The failure probabilities corresponding to $\beta$ are available as tables or as functions in up-to-date spreadsheet applications. In structural design the index $\beta$ is also referred to as the safety index or the reliability index.

Very often either R or S is constant. Then the relationship in equation (4.1) is reduced to the forms

$$\beta(t) = \frac{r - \mu[S,t]}{\sigma[S,t]} \tag{4.2}$$

or

$$\beta(t) = \frac{\mu[R,t] - s}{\sigma[R,t]} \tag{4.3}$$

where r and s are constants.

When R is constant and S is a time-related function approximated by a degradation model, the problem is called a degradation problem. Likewise when S is constant and R is a time-related function approximated by a performance model, the problem is called a performance problem.

As the means and standard deviations are dependent on time, so is index $\beta$. To obtain the distribution of service life the failure probabilities must be solved with several values of $t$ (e.g. $t = 0$, 10, 20, etc. years).

### 4.2.2 Example

The process of carbonation is studied near the surface of a

reinforced concrete structure. Failure is assumed to occur when carbonation depth exceeds the depth of reinforcement (Kasami *et al.*, 1986; Sentler, 1984). (As a result the reinforcement is depassivated and corrosion is initiated.)

In this case we have a degradation problem with a time-related carbonation process (S) and a constant concrete cover (R). Carbonation is assumed to proceed as related to the square root of time:

$$\mu(D) = K_c t^{1/2} \qquad (4.4)$$

where $\mu(D)$ = the mean of the depth of carbonation (mm),
$\qquad K_c$ = the carbonation rate factor (mm/year$^{1/2}$), and
$\qquad t$ = time (or age in years).
The carbonation rate factor depends on the strength and composition of the concrete (cf. section 8.5.3):

$$K_c = c_{env} c_{air} a (f_{ck} + 8)^b \qquad (4.5)$$

where $c_{env}$ = the environmental coefficient,
$\qquad c_{air}$ = the coefficient of air content,
$\qquad f_{ck}$ = the characteristic cubic compressive strength of concrete (MPa) and
$\qquad a, b$ = constants (depending on the binding agent).
Values of parameters and coefficients are listed in Tables 8.3–8.5. The depth of carbonation is assumed to be normally distributed. The coefficient of variation (ratio of standard deviation to mean) is assumed to be constant.

Figure 4.1 shows the process of carbonation in the concrete cover. When the mean of the carbonation depth increases with time, the standard deviation also increases, keeping the ratio of variation constant. That part of the distribution of carbonation depth that exceeds the thickness of the concrete cover shows the failure probability.

**(a) Task 1**
Find the distribution function and probability density function of service life when:

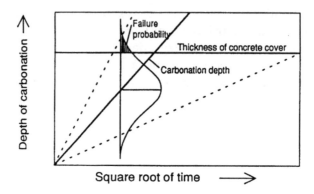

**Figure 4.1** Schematic representation of the process of carbonation in a concrete cover.

1. the structure is sheltered from rain;
2. the concrete is made of Portland cement, no air entrainment;
3. the characteristic compressive strength of concrete is 30 MPa:
4. the thickness of concrete cover is 25 mm.

In this case the thickness of concrete cover is also assumed to be a stochastic quantity. For the carbonation depth the coefficient of variance is 0.6 and the coefficient of variance of the concrete cover is 0.2.

The test index $\beta$ is calculated as follows (cf. equation (4.1):

$$\beta = \frac{25 \text{ mm} - 1800(30+8)^{-1.7}t^{1/2} \text{ mm}}{[(0.6 \times 1800(30+8)^{-1.7}t^{1/2})^2 + (0.2 \times 25)^2]^{1/2} \text{ mm}} \quad (4.6)$$

The probability distribution function and probability density function are presented in Figures 4.2 and 4.3. Note that the probability density function is not normally distributed. The probability density peaks rapidly, then slowly decreases with increasing lifetime.

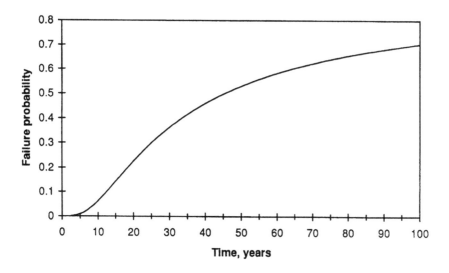

**Figure 4.2**   Probability distribution function of service life.

**(b) Task 2**
What is the required concrete cover if the target service life is 60
years with a 10% fractile (probability of failure)? The calculation
is to be performed for all binding agents in Table 8.5 and for the
characteristic compressive strengths of 30, 50 and 70 MPa.
   In the case of normal distribution, a fractile of 10%
corresponds to a $\beta$ value of 1.28. Thus the problem can be
formulated as follows: what thickness of concrete cover, $X$, will
produce a $\beta$ value of 1.28? For example, for fly ash concrete and
compressive strength 50 MPa the value of $\beta$ would be

$$\beta = \frac{X - 360(50 + 8)^{-1.2} \times 60^{1/2} \text{ mm}}{[0.6 \times 360(50 + 8)^{-1.2} \times 60^{1/2})^2 + (0.2X)^2]^{1/2} \text{ mm}} \qquad (4.7)$$

   Solution of $X$ is not too easy as this parameter is found in both
the numerator and denominator. However, the problem can be
solved unequivocally. The results are presented in Table 4.1.

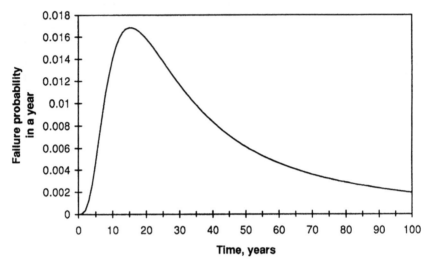

**Figure 4.3** Probability density function of service life.

**Table 4.1**  Required thickness of concrete cover (mm)

| Binder | Cover (mm) with compressive strength (MPa) of | | |
|---|---|---|---|
|  | 30 | 50 | 70 |
| Portland cement | 55 | 27 | 16 |
| Portland cement + fly ash 28% | 68 | 41 | 29 |
| Portland cement + silica fume 9% | 75 | 45 | 32 |
| Portland cement + blast furnance slag | 68 | 41 | 29 |

## 4.3 DURABILITY DESIGN WITH LOG-NORMAL SERVICE LIFE DISTRIBUTION

### 4.3.1 General

As the example above shows, although the degradation (depth of carbonation) was normally distributed around the mean, the service life distribution showed a strong decline towards short service lives. The distribution of service life in fact most often follows this pattern. The probability density peaks rapidly before decreasing gently towards zero when approaching an infinitely long service life.

For the purpose of durability design it may often be useful to make an assumption for the type of distribution of service life. It should be selected from among those with an inclined appearance such as the log-normal distribution (Figure 4.4). A log-normal distribution means that the service life is distributed normally on a logarithmic time scale.

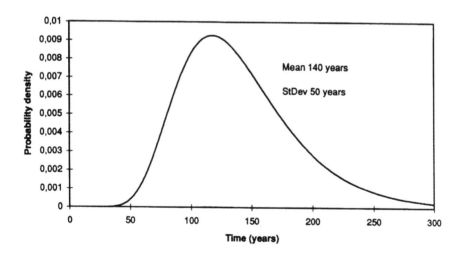

**Figure 4.4** Log-normal distribution of service life.

In a normal distribution (with linear time scale) a diminishing but still finite probability continues in both directions. This would allow negative service lives, which of course cannot be approved. The fact that negative values are impossible on a logarithmic time scale makes the assumption of a log-normal distribution for service life both suitable and plausible.

The theory of log-normal service life was first elaborated for durability problems in concrete structures by the Dutch authors Siemes, Vrouwenvelder and van den Beukel (1985). The theory also contains the necessary methods for evaluation of the standard deviation of service life.

### 4.3.2 Standard deviation of service life

The standard deviation of service life is determined using the model formula for mean service life, and the standard deviations of parameters in the formula. The model formula can generally be written as

$$t_L = t_L(x_1, x_2, \dots x_n) \tag{4.8}$$

where $t_L$ = the service life,

$x_i$ = a parameter of the material or condition, or a structural dimension ($i = 1 \dots n$).

Equation (4.8) is expected to yield the arithmetic mean of service life. $\mu(t_L)$, when the mean values of parameters $x_1, x_2, \dots, x_n$ are inserted into it.

The standard deviation is evaluated using the following formula:

$$\sigma^2(t_L) = \sum_{i=1}^{n} \left( \frac{\partial t_L}{\partial x_i} \sigma(x_i) \right)^2 \tag{4.9}$$

where $\sigma(t_L)$ = the standard deviation of service life,

$\sigma(x_i)$ = the standard deviation of parameter $x_i$,

$\dfrac{\partial t_L}{\partial x_i}$ = the partial derivative of the service life with respect to parameter $x^i$, and

$n$ = the number of variables.

### 4.3.3 Probabilities of log-normal service life distribution

If the service life distribution is log-normal with a mean value $\mu(t_L)$ and standard deviation $\sigma(t_L)$, function $Y = \ln(t_L)$ is normally distributed and its mean value and standard deviation are obtained from the following formulae:

$$\sigma^2(Y) = \ln \left[ 1 + \left( \frac{\bar{\sigma}(t_L)}{\mu(t_1}\right)^2 \right] \qquad (4.10)$$

$$\mu(Y) = \ln\mu(t_L)^{-1/2}\sigma^2(Y) \qquad (4.11)$$

The probability of the service life being shorter than a certain time $t$ is as follows:

$$P\{t_L < t\} = P\{\ln t_L < \ln t\} = P\{Y < \ln t\} = \phi(-\beta) \qquad (4.12)$$

where

$$\beta = \frac{\mu(Y) - \ln t}{\sigma(Y)} \qquad (4.13)$$

and $\phi$ is the cumulative density function of the standard normal distribution ($\mu = 0$, $\sigma = 1$).

### 4.3.4 Example

Equation (4.14) is used to calculate the mean service life with regard to corrosion of the reinforcement. The first term in the equation is the initiation time of corrosion (cf. section 8.5.3). The second term is the cracking time of corrosion (cf. section 8.5.4).

$$\mu(t_L) = \frac{C^2}{(c_{env}c_{air}a(f_{ck} + 8)^b)^2} + \frac{80C}{Dr} \qquad (4.14)$$

where $c_{env}$ = the environmental coefficient,
$\quad c_{air}$ = the coefficient of air content,
$\quad C$ = the thickness of the concrete cover (mm),
$\quad D$ = the diameter of the rebar (mm),
$\quad f_{ck}$ = the characteristic cubic strength of concrete (MPa),
$\quad r$ = the rate of corrosion in rebars, and
$\quad a, b$ = constants.
See the values of $c_{env}$, $c_{air}$, $a$ and $b$ in Tables 8.3–8.5.

In the formula for standard deviation (equation (4.9)), the standard deviations or the coefficients of variation of all parameters are needed. In Table 4.2 a choice has been made for the mean values and coefficients of variation for the problem variables. If the coefficients of variation are not known they must be evaluated.

To be able to use equation (4.9), partial differentiation with respect to every variable must be performed. If the model formula for the mean is not too complicated it can be performed by means of elementary algebra.

**Table 4.2** Problem variables and their means and coefficients of variation

| Variable | Dimension | Mean ($\mu$) | Coefficient of variation ($\sigma/\mu$) |
|---|---|---|---|
| $c_{env}$ | – | 0.5  1 | 0.5 |
| $c_{air}$ | – | 1 | 0.4 |
| $C$ | mm | 15  20  25  30  35 | 0.3 |
| $D$ | mm | 25 | 0 |
| $f_{ck}$ | MPa | 30 | 0.2 |
| $a$ | – | 1800 | 0 |
| $b$ | – | –1.7 | 0 |
| $r$ | $\mu$m/year | 16  1 | 0.5 |

The mean and the standard deviation of service life can then be calculated by inserting the mean values of the variables into the formulae. For instance, in our example, if the structures are sheltered from rain ($c_{env} = 1$, $r = 1$ $\mu$m/year) and the concrete cover is 25 mm we get

$$\mu(t_L) = 125 \text{ years}$$
$$\sigma(t_L) = 82 \text{ years}$$
$$\sigma(Y) = 0.595$$

and

$$\mu(Y) = 4.65$$

Using the tables of standard normal distribution and $\beta$ values for $t = 10$, 20, 30 etc. years, the corresponding failure probabilities can then be determined.

Figures 4.5 and 4.6 show the cumulative probability functions calculated for different thicknesses of the concrete cover. The figures show the probability of the service life being shorter than a given target service life, $t_g$. The required thickness of the concrete cover can be read from the figures as a function of the target service life and the accepted failure probability.

When using the log-normal method some requirements must be set for the model formula. One restriction is that the service life model cannot be too complicated, as it must be differentiable with respect to every parameter in it. Another requirement is that the parameters must not be related to each other; if they are, some bias is expected in the evaluation of standard deviation.

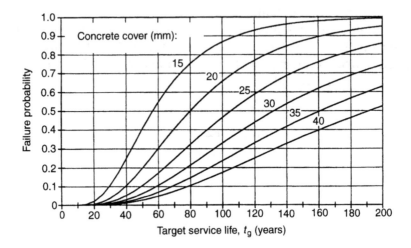

**Figure 4.5** Probability of service life being shorter than the target service life when the structures are sheltered from rain. The characteristic cubic strength of concrete is 30 MPa and the diameter of the rebar is 25 mm.

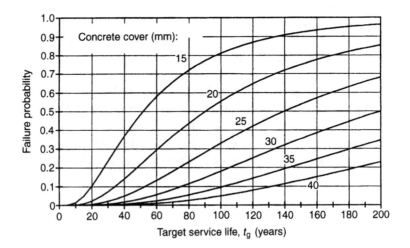

**Figure 4.6** Probability of service life being shorter than the target service life when the structures are exposed to rain. The characteristic cubic strength of concrete is 30 MPa and the diameter of the rebar is 25 mm.

# 5

# Determination of lifetime safety factors

## 5.1 THE MEANING OF LIFETIME SAFETY FACTOR

The lifetime safety factor method in the design of structures is based on the same safety principles as those applied in the stochastic methods above. With the aid of the lifetime safety factor the design problem returns to the form of deterministic design.

Figure 5.1 shows a distribution of service life and the relationships between the target service life, failure probability and mean service life. The lifetime safety factor is the relation of mean service life to the target service life.

$$\gamma_t = \frac{\mu(t_{\mathrm{L}})}{t_{\mathrm{g}}} \tag{5.1}$$

where $\gamma_t$ = the central lifetime safety factor,
$\mu(t_{\mathrm{L}})$ = the mean service life and
$t_{\mathrm{g}}$ = the target service life.
Using the lifetime safety factor, the requirement of target service life (corresponding to a maximum allowable failure probability) is converted to the requirement of mean service life. The reason for this is that the durability models available to designers show only the mean performance, or the mean degradation, or the mean service life. As designers operate with mean functions, every requirement of target service life must first be interpreted in terms of the corresponding mean service life.

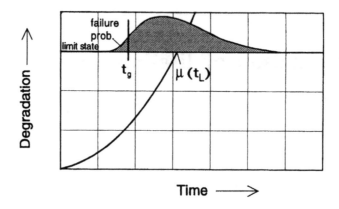

**Figure 5.1** Relationship between mean service life and target service life.

In the durability design of structures, the mean service life is often approximated by service life models which show the crossing-point of the degradation curve with the limit state of durability (Figure 5.1). The mean service life evaluated by the service life model must be greater than or equal to the **design service life**, which is the product of the lifetime safety factor and target service life.

$$\mu(t_L) \geq t_d \qquad (5.2)$$
$$t_d = \gamma_t t_g \qquad (5.3)$$

where $t_d$ is the design service life.

Note that the mean service life is not necessarily the same as the service life corresponding to 50% failure probability, which is the median service life.

The lifetime safety factor depends on the maximum allowable failure probability, the smaller the maximum allowable failure probability, the greater is the lifetime safety factor. The lifetime safety factor also depends on the form of service life distribution.

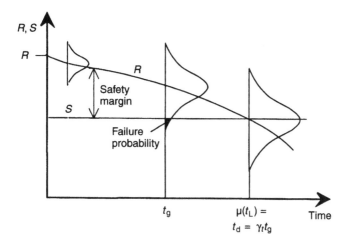

**Figure 5.2** The meaning of lifetime safety factor in a performance problem.

Figure 5.2 illustrates the meaning of lifetime safety factor when the design is carried out according to the performance principle. The curves in the figure correspond to a situation common to the design problem of load-bearing capacity. $S$ could be a constant load effect on the structure, and $R(t)$ a structural capacity which must be greater than $S$ to avoid failure. The function $R(t) - S$ is called the safety margin.

The structural performance decreases with time due to degradation of materials. The crossing-point of the $R(t)$ curve with the minimum load effect $S$ gives the mean service life which equals the design service life. If the target service life were the same as the design service life, roughly half of all structures would fall below the load requirement. To have a smaller failure probability at the target service life, the design service life must be longer than the target service life.

In general, $R(t)$ is the performance capacity of the structure. $S$, which in the case of structural performance is the load effect, is in many other applications of durability design replaced by the minimum performance capacity of the structure, $R_{min}$.

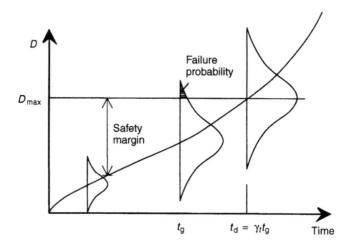

**Figure 5.3**  The meaning of lifetime safety factor in a degradation problem.

Performance behaviour can always be translated into degradation behaviour. By definition, degradation is a decrease in performance. The transformation is performed by the following substitutions:

$$R_0 - R(t) = D(t) \qquad (5.4)$$
$$R_0 - S = D_{max} \qquad (5.5a)$$

or

$$R_0 - R_{min} = D_{max} \qquad (5.5b)$$

Figure 5.3 shows the principle of design in a degradation problem. $D(t)$ is the degrading effect of environmental loading on the performance of the structure. The degradation curve crosses the maximum degradation at the design service life, which must be longer than the target service life. The range $D_{max}-D(t)$ is the safety margin.

A prerequisite for using the lifetime safety factor method is, of course, that the correct values for lifetime safety factors are known. The rest of this chapter focuses on the determination of lifetime safety factors by different methods. Practical applications of the lifetime safety factor method are presented in Chapter 6 (Structural durability design).

## 5.2 LIFETIME SAFETY FACTORS AS DETERMINED BY STOCHASTIC METHODS

### 5.2.1 Determination of lifetime safety factor with a normally distributed degradation function

Let us consider that the degradation function is of the following form:

$$\mu(D(t)) = at^n \tag{5.6}$$

where $\mu(D(t))$ = the mean of degradation,
$\quad\quad\quad a$ = the constant coefficient,
$\quad\quad\quad t$ = time, and
$\quad\quad\quad n$ = the exponent.

The exponent $n$ may in principle vary between $-\infty$ and $+\infty$. The coefficient $a$ is fixed when the mean service life is known:

$$a = \frac{D_{max}}{\mu(t_L)^n} \tag{5.7}$$

Degradation is assumed to be normally distributed around the mean. It is also assumed that the standard deviation of $D$ is proportional to the mean degradation, the coefficient of variation being constant, $\nu_D$.

Figure 5.4 shows the degradation as a function of $t^n$. The value of $\gamma_t$ can be determined as follows (Vesikari, 1995). The index $\beta$ of standard normal distribution at $t_g$ is

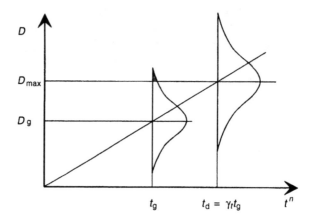

**Figure 5.4** Deterioration of lifetime safety factors with normally distributed $D$.

$$\beta = \frac{D_{max} - D_g}{\nu_D D_g} = \frac{1}{\nu_D} \left( \frac{D_{max}}{D_g} - 1 \right) \tag{5.8}$$

where $D_{max}$ = the maximum allowable degradation,

$D_g$ = the mean degradation at $t_g$, and

$\nu_D$ = the coefficient of variation of degradation.

From Figure 5.4 we get

$$\frac{D_{max}}{D_g} = \frac{(\gamma_t t_g)^n}{t_g{}^n} = \gamma_t{}^n \tag{5.9}$$

By assigning this to equation (5.8) we obtain

$$\gamma_t = (\beta \nu_D + 1)^{1/n} \tag{5.10}$$

The lifetime safety factor depends on $\beta$ (respective to the maximum allowable failure probability at $t_g$), the coefficient of variation of $D$ and the exponent $n$. Note that the lifetime safety factor is not directly dependent on $t_g$.

**Table 5.1**　Values of lifetime safety factor. Degradation curve is linear $(n = 1)$

| Failure probability (%) | $\beta$ | $\gamma_t$ with $\nu_D$ of | |
|---|---|---|---|
| | | 0.5 | 1.0 |
| 1 | 2.33 | 2.16 | 3.33 |
| 5 | 1.64 | 1.82 | 2.64 |
| 10 | 1.28 | 1.64 | 2.28 |
| 50 | 0 | 1.00 | 1.00 |

**Table 5.2**　Values of lifetime safety factor. Degradation is proportional to the square root of time $(n = \frac{1}{2})$; retarding degradation

| Failure probability (%) | $\beta$ | $\gamma_t$ with $\nu_D$ of | |
|---|---|---|---|
| | | 0.5 | 1.0 |
| 1 | 2.33 | 4.64 | 11.06 |
| 5 | 1.64 | 3.32 | 7.00 |
| 10 | 1.28 | 2.69 | 5.21 |
| 50 | 0 | 1.00 | 1.00 |

**Table 5.3**　Values of lifetime safety factor. Degradation is proportional to the square of time $(n = 2)$; accelerating degradation

| Failure probability (%) | $\beta$ | $\gamma_t$ with $\nu_D$ of | |
|---|---|---|---|
| | | 0.5 | 1.0 |
| 1 | 2.33 | 1.47 | 1.82 |
| 5 | 1.64 | 1.35 | 1.63 |
| 10 | 1.28 | 1.28 | 1.51 |
| 50 | 0 | 1.00 | 1.00 |

In the following we study the values of $\gamma_t$ as a function of $\beta, \nu_D$ and $n$. For $n$ we assume

$$n = 1 \text{ (linear degradation)}$$
$$n = 0.5 \text{ (retarding degradation)}$$
$$n = 2 \text{ (accelerating degradation)}$$

Tables 5.1–5.3 show the values of lifetime safety factor for $\beta$ values corresponding to failure probabilities of 1, 5, 10, 20 and 50%.

### 5.2.2 Determination of lifetime safety factors with a log-normally distributed service life function

In the following we assume that the distribution function of service life is log-normal. This distribution has already been introduced in section 4.3.

By inserting $t = t_g$, $\mu(t_L) = \gamma_t t_g$ and $\nu_L = \sigma(t_L)/\mu(t_L)$ into equations (4.10), (4.11) and (4.13) the following formula is obtained for $\beta$:

$$\beta = \frac{\ln(\gamma_t t_g) - \frac{1}{2}\ln(1 + \nu_L^2) - \ln(t_g)}{[\ln(1 + \nu_L^2)]^{1/2}} \qquad (5.11)$$

where $\gamma_t =$ the lifetime safety factor,
       $\beta =$ the index of standard normal distribution, and
       $\nu_L =$ the coefficient of variation of service life.
For the lifetime safety factor we obtain

$$\gamma_t = \exp\left\{\beta[\ln(1 + \nu_L^2)]^{1/2} + \frac{1}{2}\ln(1 + \nu_L^2)\right\} \qquad (5.12)$$

Thus the lifetime safety factor depends on the index $\beta$ and the coefficient of variation of service life but not directly on the length of service life. Table 5.4 shows the lifetime safety factors calculated with varying $\beta$ values.

**Table 5.4** Lifetime safety factors determined by the log-normal method

| Failure probability (%) | Lifetime safety factor, $\gamma_t$ with coefficient of variation, $\nu_L$, of | |
|---|---|---|
| | 0.5 | 1.0 |
| 1 | 3.36 | 9.81 |
| 5 | 2.43 | 5.56 |
| 10 | 2.05 | 4.11 |
| 50 | 1.12 | 1.41 |

## 5.3 DETERMINATION OF LIFETIME SAFETY FACTORS FOR STRUCTURAL DURABILITY DESIGN

### 5.3.1 Common principles

The basis of a conventional design procedure for concrete structures can be expressed as

$$R_d - S_d > 0 \qquad (5.13)$$

or

$$\Theta_d > 0 \qquad (5.14)$$

where $R_d$ = the design value of the load-bearing capacity,
$\quad S_d$ = the design value of the load and
$\quad \Theta_d$ = the safety margin calculated with the design values of load-bearing capacity and the load ($= R_d - S_d$).

The design value of the load-bearing capacity is determined by dividing the characteristic strength of concrete by a material safety factor defined for concrete, and the characteristic strength of reinforcing steel by the safety factor for steel, and then using normal design formulae for calculating the capacity. The design value of the load is obtained by multiplying the characteristic loads by their corresponding safety factors.

As the values of partial safety factors are calibrated according to the requirement of mechanical safety, the state of $R_d$ being equal to $S_d$ corresponds to the required **safety level**. 'Failure' of $R_d$ being smaller than $S_d$ does not mean a collapse or other mechanical failure, but running under the safety level requirement. Safety in terms of not falling below the safety level is called the durability safety, and is distinct from the mechanical safety. Respectively the term 'durability failure' is used distinct from the mechanical failure.

The structural durability design is very similar to normal structural design. Here the load-bearing capacity is, however, a time-dependent quantity because of the time-related material losses in concrete and steel. Possibly also the load is time-dependent. Thus the basic design formula (equation (5.14)) is rewritten as

$$\Theta_d(t_d) > 0 \qquad\qquad (5.15)$$

where $t_d$ is the design service life.

Figure 5.5 shows the reduction of the design safety margin with time. The 0-line in the figure shows the minimum value of $\Theta_d$ which fulfils the safety level requirement. The initial $\Theta_d$ must be higher than this value to be adequate at the end of the service life. Examination of the design safety margin with time shows how the probability of falling below the safety level is increased with time due to reduction of the mean of $\Theta_d$ and the increase in scatter.

Figure 5.5 also shows why it is necessary to use 'design service life' instead of 'target service life' in the design equation (5.15). If the structure were designed using the target service life, $t_g$, some 50% of structures would not meet the safety level requirement at the age of target service life. To have a smaller fractile of non-qualified structures at the target service life, a longer design service life must be used in the design formulae.

The necessary difference between the design service life and the target service life is specified by the lifetime safety factor. The method closely resembles that for dealing with the stochasticity of strengths and loads in the design formula. The target service

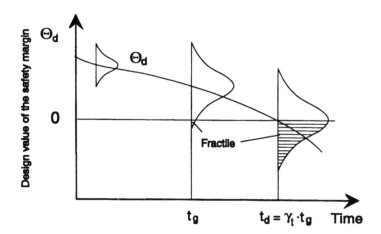

**Figure 5.5** Decrease of the design value of the safety margin, and effect of the lifetime safety factor calculated with design values of loads and load-bearing capacities.

life is now multiplied by the lifetime safety factor to obtain the design service life.

The lifetime safety factor must be calibrated in such a way that the probability of falling short of the safety level requirement because of degradation (probability of durability failure) is smaller than that required. The smaller the required probability of durability failure, the greater is the lifetime safety factor.

There are two principles for determination of the required probability of durability failure and the corresponding lifetime safety factor:

1. separated safety principle
2. combined safety principle.

According to the separated safety principle the requirement of mechanical safety and the requirement of durability safety are imposed separately.

In the principle of combined safety, the durability safety is understood as being one of several safety elements that produce the total mechanical safety. Thus the required probability of durability failure is linked to the requirement of mechanical safety.

### 5.3.2 Determination of lifetime safety factors by the separated safety principle

The separated safety principle means a separate specification of the requirements for durability safety and mechanical safety. However, in this context it has been decided to apply the requirements of mechanical safety according to Eurocode 1 also to the durability safety. By this requirement the reliability against durability failure at the end of the target service life is the same as the short-term reliability against a mechanical failure in the ordinary structural design. Some grounds for this decision are presented in section 5.3.3(d).

The required safety indexes for ordinary design defined by Eurocode 1 are as follows:

$$\text{Ultimate limit state: } \beta = 3.8 \quad (P_f = 7.2 \times 10^{-5})$$
$$\text{Serviceability limit state: } \beta = 1.5 \quad (P_f = 6.7 \times 10^{-2})$$

The ultimate limit state is associated with collapse, or with other similar forms of mechanical failure. It generally corresponds to the maximum load-carrying resistance of a structure or structural part. Serviceability limit states correspond to conditions beyond which specified service requirements for a structure or structural element are no longer met. The serviceability requirements concern (CEN, 1994)

1. functioning of the construction or parts of them;
2. the comfort of people;
3. the appearance.

If a durability failure in an ultimate state design does not lead to serious consequences, a slightly smaller value for $\beta_t$ than that specified in Eurocode 1 for mechanical safety can be allowed. On the other hand, if the consequences of a durability failure in serviceability state design are noticeable and the repair costs are high, a slightly higher value for the safety index than that stated in Eurocode 1 is required. Thus the final durability safety requirements are as follows:

1. ultimate limit state:

$\beta_t = 3.8$   (serious consequences of a durability failure)
$\beta_t = 3.1$   (no serious consequences of a durability failure)

2. serviceability limit state:

$\beta_t = 2.5$   (consequences of a durability failure are noticeable and the repair costs are high)
$\beta_t = 1.5$   (no noticeable consequences of a durability failure)

The corresponding lifetime safety factors are determined using equation (5.10). The linear mode of the degradation process is applied ($n = 1$). In Table 5.5 the lifetime safety factors are determined for different values of the coefficient of variation ($\nu_D$).

### 5.3.3 Determination of lifetime safety factors by the combined safety principle

#### (a) Principle
According to the combined safety principle the requirement of durability safety is determined on the basis of its influence on the mechanical safety. The required lifetime safety factors are determined by an analysis performed with the **characteristic values** of loads and load-bearing capacities, as the requirements of the total mechanical safety are defined with respect to characteristic values. In addition, the extra scatter due to degradation (Figure 5.5) is totalled with the ordinary scatter of the safety margin (ordinary scatter of loads and load-bearing capacity).

**Table 5.5** Values of the lifetime safety factor

| Limit state | Safety class (consequence of durability failure) | Probability of durability failure (after $t_g$), $P_f$ | Safety index (after $t_g$), $\beta_t$ | Lifetime safety factor, $\gamma_t$, with $\nu_D$ of | | |
|---|---|---|---|---|---|---|
| | | | | 0.4 | 0.6 | 0.8 |
| Ultimate limit | 1. (serious) | $7.2 \times 10^{-5}$ | 3.8 | 2.52 | 3.28 | 4.04 |
| | 2. (non-serious) | $9.7 \times 10^{-4}$ | 3.1 | 2.24 | 2.86 | 3.48 |
| Service ability limit | 1. (noticeable) | $6.2 \times 10^{-3}$ | 2.5 | 2.00 | 2.50 | 3.00 |
| | 2. (not noticeable) | $6.7 \times 10^{-2}$ | 1.5 | 1.60 | 1.90 | 2.20 |

**(b) Requirements for ordinary design**

The Eurocode requirements for the mechanical safety (cf. section 5.3.2) refer to the safety margins calculated with characteristic values of material properties and loads. This allows the graphical presentation of the safety requirements in the ultimate limit state and in the serviceability limit state as shown in Figure 5.6. In Figure 5.6, $\Theta_k$ is the mean of the 'characteristic' safety margin calculated as

$$\Theta_k = R_k - S_k \qquad (5.16)$$

where $R_k$ = the characteristic load-bearing capacity (ultimate limit state) or other structural performance (serviceability limit state) and
$S_k$ = the characteristic load.
$\sigma$ is the standard deviation of the safety margin:

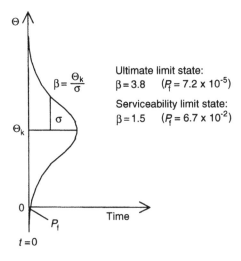

**Figure 5.6** Safety requirements for ordinary design according to Eurocode 1.

$$\sigma = \sigma(\Theta_k) = \frac{\Theta_k}{\beta} \tag{5.17}$$

## (c) Derivation of requirements for durability design

In durability design the mechanical safety index after the service life must be at least $\beta_m$, which as a rule can be the same as that defined for the ultimate limit state in ordinary design. However, in cases where no serious consequences of a structural failure are expected, the safety index $\beta_m$ could be a little smaller than that specified for an ordinary designed structure. The reason is that in ordinary (traditional) design the safety is only specified at the start of service life. As a result of degradation the actual safety is bound to fall under the specified level towards the end of service life. Thus the dimensions of a durability designed structure with a lowered safety index $\beta_m$ are probably more consistent with those of a traditionally designed structure. However, the safety of a durability designed structure is under control throughout the service life.

For the above reason, the same safety classification presented in section 5.3.2 is now applied to mechanical safety:

1. ultimate limit state:

$\beta_m = 3.8$    (serious consequences of a durability failure)
$\beta_m = 3.1$    (no serious consequences of a durability failure)

2. serviceability limit state:

$\beta_m = 2.5$    (consequences of a durability failure are noticeable and the repair costs are high)
$\beta_m = 1.5$    (no noticeable consequences of a durability failure)

In the following analysis it is assumed that the loads and material strengths remain constant throughout the service life of the structure, and their effects on the safety margin are determined in basically the same way as in ordinary design. The degradation of materials is considered to affect the cross-sectional dimensions of materials (cf. Chapter 6) by both reducing the means and increasing the standard deviations, with corresponding effects on the safety margin.

The time (or degradation)-related extra scatter is taken into account by the lifetime safety factor, $\gamma_t(\gamma_t > 1)$, in such a way that the determinative time in the design is the design time, $t_d$, which is obtained by multiplying the target service life, $t_g$, by the lifetime safety factor. This means that, when expressing the basic formulae of durability design for the time $t = t_d$, the time-related extra scatter is not taken into account as the effect of the scatter is included in the safety factor, $\gamma_t$. The safety factor is determined in such a way that the safety index, $\beta$, at the time $t = t_d$, when only the reduction of the mean is taken into account, is the same as that for the time $t = t_g$, when the extra scatter due to degradation is also taken into account (Vesikari, 1995).

In the following analysis the characteristic safety margin $\Theta(t)$, defined as in equation (5.16), is expressed as

$$\Theta(t) = \Theta_o - \Delta\Theta(t) \tag{5.18}$$

where $\Theta_o$ = the mean of the characteristic safety margin at $t = 0$
            and
$\Delta\Theta$ = the decrease in the characteristic safety margin due
            to degradation.
$\sigma_o$ is the standard deviation of $\Theta_o$ (constant) and $\sigma_t$ is the
standard deviation of $\Delta\Theta$ (due to degradation, corresponds to
the scatter in Figure 5.5). We then write

$$\sigma_{\text{tot}} = (\sigma_o{}^2 + \sigma_t{}^2)^{1/2} \qquad (5.19)$$

$$\alpha_t = \frac{\sigma_t}{\sigma_{\text{tot}}} \qquad (5.20)$$

where $\sigma_{\text{tot}}$ = the total standard deviation $\Theta$ and
            $\alpha_t$ = the ratio of standard deviation due to degradation
            to the total standard deviation.
We also denote:
$\Theta_m$ = the minimum value of $\Theta$ corresponding to the safety
            index $\beta_m$ (corresponds also to the 0-line in Figure 5.5).
$\beta_t$ = the partial safety index with regard to degradation
            (safety index of durability failure).
In the following examination the lower index $t$ refers to the
time $t = t_g$. The mean of $\Delta\Theta$ is assumed to increase
proportionally to the function $t^n$, where $n$ is the exponent of
time and $\sigma_t$ is directly proportional to $\Delta\Theta$.
Figure 5.7 shows the diminishing with time of safety margin
$\Theta$. The mechanical failure probability is described by that part of
the distributions which has fallen below the limit $\Theta = 0$. The
failure probability is influenced both by the decrease in the mean
of $\Theta$ and the increase of standard deviation with time.
According to Figure 5.7 the safety index at $t = t_d$, when the
time-related scatter is not taken into account, is

$$\beta_d = \frac{\Theta_m}{\sigma_o} \qquad (5.21)$$

According to the principle mentioned above, the safety index $\beta_d$
must be equal to the safety index $\beta_g$ at $t = t_g$ when the time-
related scatter is also included. At $t = t_g$ the mean of $\Theta$ deviates
from $\Theta_m$ by the amount $\beta_t\sigma_t$.

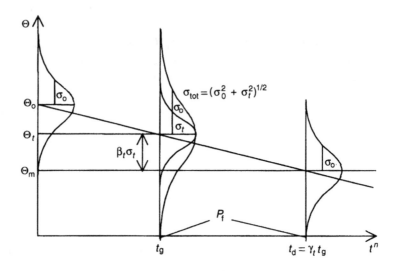

**Figure 5.7**   Decrease of the safety margin with time.

$$\beta_{\mathrm{g}} = \frac{\Theta_{\mathrm{m}} + \beta_t \sigma_t}{\sigma_{\mathrm{tot}}} \tag{5.22}$$

By setting $\beta_{\mathrm{d}}$ equal to $\beta_{\mathrm{g}}$ and applying equation (5.20) we obtain

$$\frac{\Theta_{\mathrm{m}}}{\sigma_{\mathrm{o}}} = \frac{\Theta_{\mathrm{m}}(1 - \alpha_t^2)^{1/2}}{\sigma_{\mathrm{o}}} + \beta_t \alpha_t$$

Further applying the requirement $\beta_{\mathrm{d}} = \Theta_{\mathrm{m}}/\sigma_{\mathrm{o}} = \beta_{\mathrm{m}}$ gives

$$\beta_{\mathrm{m}} = \beta_t \frac{\alpha_t}{1 - (1 - \alpha_t^2)^{1/2}} \tag{5.23}$$

Equation (5.23) describes how the relation of the total mechanical safety requirement ($\beta_{\mathrm{m}}$) to the durability safety requirement ($\beta_t$) depends on the ratio $\alpha_t$.

$\alpha_t$ can be determined as follows. According to Figure 5.7 we get for $\Theta_t$:

$$\Theta_t = \Theta_o(1 - kt^n) \tag{5.24}$$

where $k$ is the reduction coefficient of $\Theta$. Thus we get

$$\Delta\Theta = \Theta_o kt^n \tag{5.25}$$

and

$$\sigma_t = \nu_D \Theta_o kt^n \tag{5.26}$$

where $\nu_D$ is the coefficient of variation of the decrease of $\Theta$. $\sigma_o$ can be expressed as

$$\sigma_o = \Theta_m/\beta_m \tag{5.27}$$

For $k$ we can denote

$$k = \frac{\Theta_o - \Theta_m}{\Theta_o t_d^n} = \frac{m}{(\gamma_t t_g)^n} \tag{5.28}$$

where the quantity $m$ is the relative reduction of $\Theta$ during the interval $0 \rightarrow t_d$:

$$m = \frac{\Theta_o - \Theta_m}{\Theta_o} = 1 - \frac{\Theta_m}{\Theta_o} \tag{5.29}$$

From equations (5.19) and (5.20) we can then derive for $\alpha_t$ at $t = t_g$:

$$\alpha_t = \frac{m\nu_D \Theta_o/\gamma_t^n}{[(\Theta_m/\beta_m)^2 + (m\nu_D \Theta_o/\gamma_t^n)^2]^{1/2}} \tag{5.30}$$

By dividing both the numerator and the denominator by $\Theta_m/\beta_m$ and taking into account that $\Theta_o/\beta_m = 1/(1 - m)$ we get finally

$$\alpha_t = \frac{r_t}{(1 + r_t^2)^{1/2}} \tag{5.31}$$

where

$$r_t = \frac{\nu_D m \beta_m}{(1 - m)\gamma_t^n} \tag{5.32}$$

Let us then examine the safety index $\beta_t$. From Figure 5.7 we obtain

$$\beta_t \sigma_t = \Theta_t - \Theta_m$$

from which we can derive

$$\beta_t = \frac{\Theta_t - \Theta_m}{\nu_D(\Theta_o - \Theta_t)} = \frac{1}{\nu_D}\left(\frac{\Theta_o - \Theta_m}{\Theta_o - \Theta_t} - 1\right)$$

On the other hand we can see from Figure 5.7:

$$\frac{\Theta_o - \Theta_m}{\Theta_o - \Theta_t} = \frac{(\gamma_t t_g)^n}{t_g^n} = \gamma_t^n$$

Assigning this to the formula for $\beta_t$ gives

$$\beta_t = \frac{(\gamma_t^n - 1)}{\nu_D} \tag{5.33}$$

The result is the same as equation (5.10).

By assigning equations (5.31) and (5.33) to equation (5.23) the values of $\gamma_t$ can be determined as a function of $\nu_D, n$ and $m$. The corresponding values for $\beta_t$ and $\alpha_t$ can then be obtained from equations (5.33) and (5.31).

**(d) Results and application**
In Table 5.6 the values of $\gamma_t$, $\beta_t$ and $\alpha_t$ have been determined for varying values of $m$. The safety index $\beta_m$ is 3.8. For the other parameters we assume

**Table 5.6**   Values of $\gamma_t$, $\beta_t$ and $\alpha_t$. $\beta_m = 3.8$

| $m$ | $\gamma_t$ | $\beta_t$ | $\alpha_t$ |
|---|---|---|---|
| 0.001 | 1.00 | 0.00 | 0.00 |
| 0.1 | 1.23 | 0.39 | 0.20 |
| 0.2 | 1.44 | 0.73 | 0.37 |
| 0.3 | 1.63 | 1.05 | 0.51 |
| 0.4 | 1.83 | 1.38 | 0.64 |
| 0.5 | 2.02 | 1.71 | 0.75 |
| 0.6 | 2.23 | 2.06 | 0.84 |
| 0.7 | 2.46 | 2.43 | 0.91 |
| 0.8 | 2.70 | 2.84 | 0.96 |
| 0.9 | 2.97 | 3.29 | 0.99 |
| 0.999 | 3.28 | 3.79 | 1.00 |

$$\nu_D = 0.6$$
$$n = 1$$

The corresponding values for $\beta_m = 3.1$ are given in Table 5.7 (other parameters as above).

Tables 5.8 and 5.9 apply to serviceability limit state design. Values of $\gamma_t$, $\beta_t$ and $\alpha_t$ are given for $\beta_m = 2.5$ and 1.5. $\nu_D$ and $n$ are the same as above.

The required value for $\beta_t$ and $\gamma_t$ depends on $m$ (the relative decrease of $\Theta$ during the time interval $0 \rightarrow t_d$). In principle $\gamma_t$ could be any value in Tables 5.6–5.9, if it is consistent with the real $m$. The real $m$ depends on the amount of degradation during the service life and should be checked by calculations after the design. The real $m$ must not be greater than that respective to $\gamma_t$ which was applied in the design.

In practical design a standard assumption for $m$ is 0.7 corresponding to the values of $\gamma_t$ of 2.46 for $\beta_m = 3.8$ and 2.15 for $\beta_m = 3.1$ respectively. After design of the structure the real $m$ is checked to ensure that it does not exceed 0.7.

**Table 5.7**  Values of $\gamma_t$, $\beta_t$ and $\alpha_t$. $> \beta_m = 3.1$

| $m$ | $\gamma_t$ | $\beta_t$ | $\alpha_t$ |
|---|---|---|---|
| 0.001 | 1.00 | 0.00 | 0.00 |
| 0.1 | 1.16 | 0.27 | 0.17 |
| 0.2 | 1.32 | 0.53 | 0.33 |
| 0.3 | 1.47 | 0.79 | 0.48 |
| 0.4 | 1.63 | 1.05 | 0.61 |
| 0.5 | 1.79 | 1.32 | 0.72 |
| 0.6 | 1.97 | 1.61 | 0.82 |
| 0.7 | 2.15 | 1.92 | 0.90 |
| 0.8 | 2.36 | 2.27 | 0.95 |
| 0.9 | 2.59 | 2.66 | 0.99 |
| 0.999 | 2.86 | 3.10 | 1.00 |

**Table 5.8**  Values of $\gamma_t$, $\beta_t$ and $\alpha_t$. $\beta_m = 2.5$

| $m$ | $\gamma_t$ | $\beta_t$ | $\alpha_t$ |
|---|---|---|---|
| 0.001 | 1.00 | 0.00 | 0.00 |
| 0.1 | 1.11 | 0.19 | 0.15 |
| 0.2 | 1.22 | 0.37 | 0.29 |
| 0.3 | 1.34 | 0.57 | 0.43 |
| 0.4 | 1.46 | 0.77 | 0.56 |
| 0.5 | 1.59 | 0.99 | 0.69 |
| 0.6 | 1.74 | 1.23 | 0.79 |
| 0.7 | 1.89 | 1.49 | 0.88 |
| 0.8 | 2.07 | 1.78 | 0.95 |
| 0.9 | 2.27 | 2.11 | 0.99 |
| 0.999 | 2.50 | 2.50 | 1.00 |

**Table 5.9** Values of $\gamma_t$, $\beta_t$ and $\alpha_t$. $\beta_m = 1.5$

| $m$ | $\gamma_t$ | $\beta_t$ | $\alpha_t$ |
|---|---|---|---|
| 0.001 | 1.00 | 0.00 | 0.00 |
| 0.1 | 1.04 | 0.07 | 0.10 |
| 0.2 | 1.09 | 0.15 | 0.20 |
| 0.3 | 1.15 | 0.25 | 0.32 |
| 0.4 | 1.21 | 0.35 | 0.44 |
| 0.5 | 1.28 | 0.47 | 0.57 |
| 0.6 | 1.37 | 0.62 | 0.70 |
| 0.7 | 1.47 | 0.78 | 0.82 |
| 0.8 | 1.59 | 0.98 | 0.92 |
| 0.9 | 1.73 | 1.21 | 0.98 |
| 0.999 | 1.90 | 1.50 | 1.00 |

The value 0.7 is based on the assumption that the safety margin decreases roughly linearly in this range (corresponding approximately to 0.5 in the relative reduction of load-bearing capacity $R$, cf. section 6.1.2). During the time interval $0 \rightarrow t_g$ the relative decrease of $\Theta$ is of course much less. Also the difference in capacities of an ordinarily designed and a durability designed structure would be too great if $m$ were greater than 0.7.

If $m$ turns out to be greater than 0.7 the designer should check whether there is any way to reduce it by changing the dimensions or material specifications or by reducing the rates of degradation. If not, the required target service life may not be realistic in the environment concerned and should be shortened. Then the designer should propose a new requirement for the target service life.

From Tables 5.6–5.9 it is seen that the values of $\beta_t$ and $\gamma_t$ respective to the maximum value of $m$ $(m = 1)$ are equal to the requirements of durability safety in section 5.3.2. From this one can conclude that when applying the separated safety principle with the requirements presented in Table 5.5 no check for $m$ is required after the design.

### 5.3.4 Partial safety factors of loads and materials in durability design

If $\beta_m$ is 3.8 the partial safety factors defined by Eurocode 1 (CEN, 1994) and CEB/FIP model code (CEB, 1988) are used in durability design.
Loads:

$$\gamma_g = 1.35 \quad \text{(dead loads)}$$
$$\gamma_p = 1.5 \quad \text{(variable loads)}$$

Materials:

$$\gamma_c = 1.5 \quad \text{(concrete)}$$
$$\gamma_s = 1.15 \quad \text{(steel)}$$

If $\beta_m$ is 3.1 the following partial safety factors for loads and materials are used:
Loads:

$$\gamma_g = 1.30 \quad \text{(dead loads)}$$
$$\gamma_p = 1.38 \quad \text{(variable loads)}$$

Materials:

$$\gamma_c = 1.40 \quad \text{(concrete)}$$
$$\gamma_s = 1.13 \quad \text{(steel)}$$

The partial safety factors for $\beta_m = 3.1$ have been determined according to Eurocode 1, Annex A (CEN, 1994) using equations (5.34)–(5.36). For dead loads the normal distribution and for variable loads the Gumbel distribution has been assumed. The safety factors of materials have been determined assuming log-normal distribution.
Normal distribution:

$$\gamma = (1 + \alpha\beta V_x)\xi \tag{5.34}$$

Gumbel distribution:

$$\gamma - \left(1 + V_x(0.577 + \ln\{-\ln\phi(\alpha\beta)\})\frac{\sqrt{6}}{\pi}\right)\xi \qquad (5.35)$$

Log-normal distribution:

$$\gamma = \exp(\alpha\beta V_x) \qquad (5.36)$$

In equations (5.34)–(5.36)
$\alpha =$ the sensitivity coefficient,
$\beta =$ safety index,
$V_x =$ coefficient of variation,
$\xi =$ extra safety coefficient,
$\phi =$ the distribution function normal distribution.
The following values have been used in the calculations:

$$\xi = 1.05$$
$$\alpha = 0.7 \text{ for loads}$$
$$\alpha = 0.8 \text{ for material strengths}$$

$V_x$ has been determined backwards by applying $\beta = 3.8$ and the corresponding values of $\gamma$. The new values of $\gamma$ have then been determined by applying the calculated values for $V_x$ and $\beta = 3.1$.

If $\beta_m$ is smaller than 3.8 (the principal safety requirement according to Eurocode 1) it is necessary to check that the safety margin $\Theta_o$ and the bearing capacity $R_o$ are not smaller than those obtained by normal design. If they are, the dimensions obtained by normal design are determinative. This check is to ensure that the final safety of the structure is never less than that of a normally designed structure.

# 6

# Structural durability design

## 6.1 FORMULATION OF LOAD-BEARING CAPACITY WITH TIME

### 6.1.1 Principles

The following degradation factors may have long-term effects on the load-bearing capacity of concrete structures:

1. corrosion due to chloride penetration
2. corrosion due to carbonation
3. mechanical abrasion
4. salt weathering
5. surface deterioration
6. frost attack.

Additionally there exist some internal degradation processes, such as alkaline–aggregate reaction, that are not primarily caused by environmental stresses. Such degradation problems are not treated here as they can be solved by a proper selection of raw materials and an appropriate design of concrete mix.

Degradation factors affect either the concrete or the steel or both. Usually degradation takes place on the surface zone of concrete or steel, gradually destroying the material. The main structural effects of degradation in concrete and steel are the following (Andrade *et al.*, 1989):

1. loss of concrete leading to reduced cross-sectional area of the concrete;
2. corrosion of reinforcement leading to reduced cross-sectional area of steel bars. Corrosion may occur:
   (a) at cracks;
   (b) at all steel surfaces, assuming that the corrosion products are able to leach out through the pores of the concrete (general corrosion in wet conditions);
3. splitting and spalling of the concrete cover due to general corrosion of reinforcement, leading to a reduced cross-sectional area of the concrete and a reduced bond between concrete and reinforcement.

The capacity of structures can in principle be studied at three different levels (Andrade *et al.*, 1989). At the first level only the cross-section of a structure is studied against different action effects such as bending moment, shear force, axial force, etc. The second level introduces a deterioration model of isolated structural elements such as simply supported beams or columns, taking into account deformations, sliding and buckling of reinforcement, etc. Finally, at the third level the whole structure is considered, taking into account the possible redistribution of action effects provided this is allowed by the remaining ductility of materials.

In the following only a first-level study is introduced. The load-bearing capacity of structures is studied by the design formulae for cross-sections of structures. The design formulae are supplied with time-dependent degradation models of concrete and steel. This is demonstrated below using several examples.

### 6.1.2 Examples of the calculation of degradation processes

The purpose is to show how the degradation of materials is taken into account in the formulation of load-bearing capacity, and to provide insight into the way in which the load-bearing capacity is reduced with time.

Cross-sections of a simple axially loaded column and a bent beam are studied. Degradation in both concrete and steel is

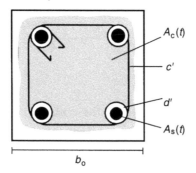

**Figure 6.1** Cross-section of a column.

assumed, leading to a reduction in the concrete and in steel cross-sections. No splitting or spalling around the corroded steel bars is assumed.

**(a) Column**
The load-bearing capacity of an axially loaded square concrete cross-section reinforced with rebars at its corners is (Figure 6.1)

$$R_d = A_c(t)\frac{f_{ck}}{\gamma_c} + A_s(t)\frac{f_y}{\gamma_s} \tag{6.1}$$

where

$R_d$ = the compressive capacity of the cross-section,
$A_c$ = the cross-sectional area of the concrete,
$A_s$ = the cross-sectional area of the steel,
$f_{ck}$ = the characteristic compressive strength of concrete,
$f_y$ = the nominal yield strength of steel,
$\gamma_c$ = the partial safety factor of concrete, and
$\gamma_s$ = the partial safety factor of steel.

The cross-sectional areas of concrete and steel are

$$A_c(t) = (b_o - 2c'(t))^2 \tag{6.2}$$

$$A_s(t) = \frac{4\pi(D_o - 2d'(t))^2}{4} \tag{6.3}$$

where

$b_o$ = the original width of the column,
$D_o$ = the original diameter of steel bars,
$c'$ = a degradation model of concrete expressing the depth of deterioration of concrete, and
$d'$ = a degradation model of steel expressing the depth of corrosion in reinforcement.

If $c'$ is small compared to $b_o$ and $d'$ is small compared to $D_o$, the following approximation of equation (6.1) can be written:

$$R_d \approx R_{do} - 4\left(b_o c'(t)\frac{f_{ck}}{\gamma_c}\right) + D_o d'(t)\frac{f_y}{\gamma_s} \tag{6.4}$$

Equation (6.4) shows that at least at the start of service life (when $c'$ and $d'$ are relatively small), the degradation of $R_d$ follows the type of degradation in concrete and steel. If $c'$ and $d'$ are linear with respect to time, the capacity of the structure will also decrease linearly with time. If $c'$ and $d'$ are accelerating or retarding, the reduction in $R_d$ will show the same tendency.

Figure 6.2 shows the reduction in $A_c, A_s$ and $R_d$ expressed in per cent (according to equations (6.1)–(6.3)) for a typical axially loaded column, when $c'$ and $d'$ are assumed to be linear with time. The calculations have been done for the following dimensions and material specifications:

**Column**

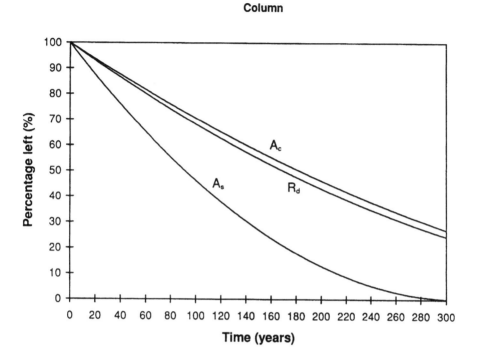

**Figure 6.2** Reductions in material cross-sections and compressive capacity of a column.

$$b_o = 500 \text{ mm}$$
$$D_o = 25 \text{ mm}$$
$$f_{ck} = 40 \text{ MPa}$$
$$f_{yk} = 400 \text{ MPa}$$
$$\gamma_c = 1.5$$
$$\gamma_s = 1.15$$
$$c'(t) = 0.4 \text{ mm/year}$$
$$d'(t) = 0.04 \text{ mm/year}$$

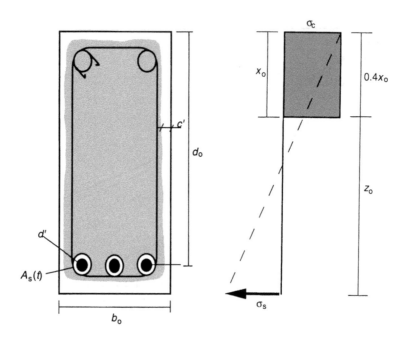

**Figure 6.3** Cross-section of a beam.

The degradation of $R_d$ seems to be roughly linear with time and closely follows the reduction of $A_c$. The corrosion of steel has only a small effect on the total capacity of the column. Even if the entire steel area were to disappear, the beam would retain some load-bearing capacity.

**(b) Beam**

The load-bearing capacity of a beam is determined by the following formulae (Figure 6.3):

$$R_{ds} = A_s(t)z(t)\frac{f_y}{\gamma_s} \quad \text{(the stress of steel is decisive)} \tag{6.5}$$

$$R_{dc} = b(t)x(t)z(t)\frac{f_c}{2\gamma_c} \quad \text{(the stress of concrete is decisive)} \tag{6.6}$$

$$x(t) = d(t)\mu(t)n\left[-1 + \left(1 + \frac{2}{(\mu(t)n)}\right)^{1/2}\right] \tag{6.7}$$

$$z(t) = d(t) - \frac{0.8x(t)}{2} \tag{6.8}$$

$$n = \frac{E_s}{E_c} \tag{6.9}$$

$$\mu(t) = \frac{A_s(t)}{b(t)d(t)} = \frac{N_s\pi(D_o - d'(t))^2/4}{(b_o - 2c'(t))(d_o - c'(t))} \tag{6.10}$$

where $R_{ds}$ = the bending capacity of the beam when the tensile
stress of steel is decisive,

$R_{dc}$ = the bending capacity of the beam when the
compressive strength of concrete is decisive,

$A_s$ = the cross-sectional area of the steel,

$d$ = the effective height of the beam,

$x$ = the distance of the neutral axis from the top surface
of the structure,

$z$ = the internal lever arm of the moment,

$E_s$ = the modulus of elasticity of steel,

$E_c$ = the modulus of elasticity of concrete,

$\mu$ = the geometric area of steel,

$c'$ = a degradation model expressing the depth of
deterioration in concrete,

$d'$ = a degradation model expressing the depth of
corrosion in reinforcement and

$N_s$ = the number of steel bars.

Beam

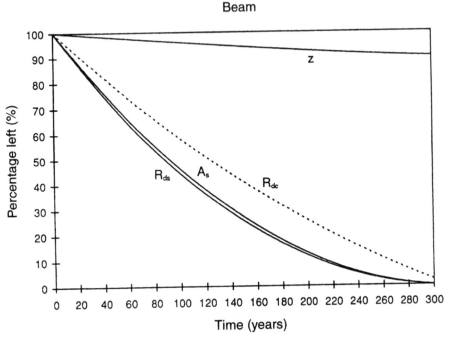

**Figure 6.4**   Reduction in the bending capacity of a beam.

The reduction in $A_s, z, R_{ds}$ and $R_{dc}$ expressed in per cent for a typical beam with constant degradation rate $c'$ and $d'$ is shown in Figure 6.4. The following dimensions and material specifications have been used in the calculations:

$$b_o = 400 \text{ mm}$$
$$d_o = 700 \text{ mm}$$
$$D_o = 25 \text{ mm}$$
$$N_s = 3$$
$$f_c = 40 \text{ MPa}$$
$$f_y = 400 \text{ MPa}$$
$$E_s = 200\,000 \text{ N/mm}^2$$
$$E_c = 9500(f_{ck} + 8)^{1/3} \text{ N/mm}^2$$
$$\gamma_c = 1.5$$
$$\gamma_s = 1.15$$

The reduction in $R_{ds}$ seems to follow closely the reduction in cross-sectional area of the steel. This is because the reduction of $z$ with time seems very small and is relatively smaller than that of $d$, as $d$ is diminishing as well.

The absolute value of $R_{ds}$ is smaller than that of $R_{dc}$. Thus $R_{ds}$ is decisive in the present study. However, the $R_{dc}$ curve is also presented in Figure 6.4, showing an almost linear reduction tendency.

## 6.2 PROPOSED PROCEDURES FOR STRUCTURAL DURABILITY DESIGN

The proposed design procedure includes the following phases:

1. ordinary mechanical design
2. durability design
3. final design.

Two methods are proposed for phase 3 (final design):

1. separated design method comprising a simple integration of the results of phases 1 and 2;
2. combined design method where the mechanical design is newly performed as $R_d(t_d) > S_d(t_d)$, applying the results of phase 2. In this case phase 1 serves mainly as a reference but may be determinative in some cases.

A flow chart of the design procedure is presented in Figure 6.5.

### 6.2.1 Ordinary mechanical design

Ordinary mechanical design is performed using conventional design methods. Its purpose is to determine the preliminary dimensions for the structure.

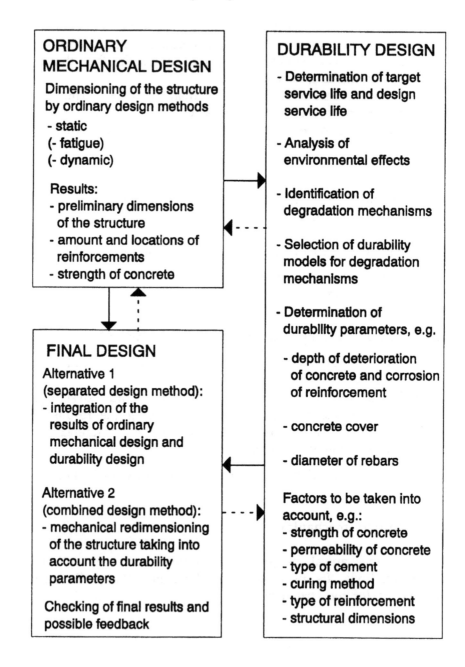

**Figure 6.5**  Flow chart of the durability design procedure.

**Table 6.1** Classification of the target service life (design working life) (source: CEN, *ENV 1991–1. Euro 1*, published by CEN, 1994)

| Class | Target service life (years) | Example |
|---|---|---|
| 1 | [1–5] | Temporary structures |
| 2 | [25] | Replaceable structural parts, e.g. gantry girders, bearings |
| 3 | [50] | Building structures and other common structures |
| 4 | [100] | Monumental building structures, bridges and other civil engineering structures |

## 6.2.2 Durability design

### (a) Durability design procedure

The aim of durability design is to improve the control of durability over the required service life. The durability design procedure is the following:

1. specification of the target service life and design service life;
2. analysis of environmental effects;
3. identification of durability factors and degradation mechanisms;
4. selection of a durability calculation model for each degradation mechanism;
5. calculation of durability parameters using available calculation models;
6. possible updating of the calculations of the ordinary mechanical design (e.g. own weight of structures);
7. transfer of the durability parameters into the final design.

### (b) Specification of target service life and design service life

The target service life is defined corresponding to the

requirements given in common regulations, codes and standards in addition to possible special requirements of the client.

Typical classes of service life are 10, 25, 50, 75, 100, etc. years. Eurocode 1 (CEN, 1994) presents the classification for the target service life given in Table 6.1 (Eurocode 1 uses here the term 'design working life' which can, however, be interpreted as the minimum requirement for service life). Another classification is presented in the standard BS 7543 (BSI, 1992).

The proposed safety classification of durability design is presented in Table 6.2. The required lifetime, load and material safety factors are also given in the table.

The design service life is determined by equation (6.11):

$$t_d = \gamma_t t_g \qquad\qquad (6.11)$$

where

$t_d$ = the design service life,
$\gamma_t$ = the lifetime safety factor, and
$t_g$ = the target service life.

### (c) Analysis of environmental effects
The analysis of environmental effects includes identification of the climatic conditions such as temperature and moisture variations, rain, condensation of moisture, freezing, solar radiation and aerial pollution, and the identification of geological conditions such as the location of ground water, possible contact with sea water, contamination of the soil by aggressive agents like sulphates and chlorides. Man-made actions such as salting of roads, abrasion by traffic, etc. must also be identified.

### (d) Identification of degradation factors and degradation mechanisms
Based on the environmental effect analysis the designer identifies the degradation factors to which the structure will most likely be subjected. The list of degradation factors in Table 7.1 may be used at this stage.

If the structures are not fully protected from the environment, some kind of degradation process is usually assumed to take place in both the concrete and the reinforcement.

### (e) Selection of durability models for each degradation mechanism

A designer must determine which degradation factors are decisive for service life. Preliminary evaluations of rates of degradation for different factors may be necessary. The models presented in Chapter 8 may be applied in these evaluations. The same design procedure and principles can be applied for all types of deterioration.

In concrete structures exposed to normal outdoor conditions the effects of degradation mechanisms can be classified into the following structural deterioration mechanisms.

1. Corrosion of reinforcement at cracks in the concrete, causing a reduction in the cross-sectional area of steel bars. The models and corrosion rate expectations presented in section 8.5 can be used in the calculations.
2. Surface deterioration or frost attack, causing a reduction in the cross-sectional area of concrete. The models presented in section 8.3 for surface deterioration or section 8.2 for frost attack can be used in the calculations.

Service life models for general corrosion presented in section 8.5 are used in the evaluation of the cracking time of concrete covers.

### (f) Calculation of durability parameters through calculation models

The depth of corrosion is determined using the design service life, $t_d$, as time. The diameter of hoops (stirrups) and other possible non-load-bearing rebars is then designed. The minimum initial diameter of rebars, $D_{omin}$, is twice the depth of corrosion added to the minimum final (after the service life) diameter of the rebar. The minimum final diameter is specified by the designer $(\geq 0)$. The initial diameter must then be chosen to be consistent with the standards $(D_o \geq D_{omin})$.

The depth of deterioration of concrete is evaluated using the

**Table 6.2** Safety classification of durability design and the corresponding safety factors

| Limit state | Safety class of durability design | Separated design | | Combined design[a] | |
|---|---|---|---|---|---|
| | | Lifetime safety factor $\gamma_t$[b] | Load and material safety factors | Lifetime safety factor $\gamma_t$[c] | Load and material safety factors |
| Ultimate limit state | 1. Serious social, economic or ecological consequences of a mechanical failure | 3.3 | Normal[d] | 2.5 | Normal[d] $\gamma_g = 1.3$[e] $\gamma_p = 1.38$ $\gamma_c = 1.4$ $\gamma_s = 1.13$ |
| | 2. Consequences of a mechanical failure are not serious | 2.9 | Normal[d] | 2.2 | |
| Serviceability limit state | 1. Noticeable consequences and considerable repair costs | 2.5 | – | 1.9 | – |
| | 2. Non-noticeable consequences and repair costs | 1.9 | – | 1.5 | – |

**Table 6.2** Safety classification of durability design and the corresponding safety factors (continued)

[a] Check for $m$ (relative reduction of the safety margin during $0 \to t_d$) $\leq 0.7$ is required.

[b] Cf. Table 5.5.

[c] Cf. Tables 5.6–5.9.

[d] Load and material safety factors specified for ordinary mechanical design are used.

[e] Reduced values of load and material safety factors may be used in the durability design (cf. section 5.3.4). However, the safety at the start of service life ($t = 0$) must be at least the same as that required in ordinary design.

design service life, $t_d$, as time. The minimum thickness of the concrete cover, $C_{omin}$, is dimensioned by adding the depth of deterioration of concrete to the minimum final cover thickness (after the service life). The minimum final thickness of concrete cover is specified by the designer ($\geq 0$). The initial thickness of concrete cover, $C_o$, is then selected by possibly rounding off upwards taking into account the requirements of the codes.

In addition, a check for general corrosion must be performed. The purpose is to ensure that no cracking or spalling of the concrete cover that would violate the reinforcement bond can take place during the service life. This can be done by applying the service life models for general corrosion with the earlier dimensioned concrete cover and diameter of steel bars as parameters. The determinative rebars with respect to cracking and spalling are normally the hoops.

### (g) Possible updating of calculations in ordinary mechanical design

Some durability parameters may influence the mechanical design. Such an interactive influence would be an increase in concrete dimensions, which increases the dead load of horizontal structures, thus increasing the load effects on both the horizontal and vertical structures. This phase is not necessary when the combined method is used.

### (h) Transfer of durability parameters to the final design

The parameters of the durability design are listed and transferred to the final design phase for use in the final dimensioning of the structure.

### 6.2.3 Final design

### (a) Separated design method

In the separated design method the mechanical design and the durability design are separated. The ordinary structural design (phase 1) produces the mechanical safety and serviceability parameters whereas the durability design (phase 2) produces the durability parameters. Both of these groups of parameters are then combined in the final design of the structure.

The possible interaction between mechanical design parameters and durability design parameters is not normally taken into account. In some cases there may exist some interaction and need for updating the mechanical design parameters as discussed in section 6.2.2(g). A fairly common case might be an increase in strength of concrete due to durability, which again influences the load-bearing capacity and reduces the dimensions especially in the case of compressed structures.

The dimensions of ordinary design are assumed to be those existing at the end of service life. Thus these dimensions must be increased by the amounts corresponding to the depth of deterioration and corrosion during service life.

The depth of deterioration of concrete is added to the structural dimensions obtained in ordinary design. If two opposite sides of the structure are exposed to degradation the dimension in ordinary design is increased by twice the depth of deterioration. The final dimension may then be rounded off upwards.

The diameter of rebars in the ordinary design of the structure is increased by twice the depth of corrosion. The final diameter must then be rounded off upwards to be consistent with standards.

No check for $m$ (relative reduction of the safety margin during $0 \rightarrow t_d$) is required.

**(b) Combined design method**
In the combined design method the mechanical design is newly performed, taking into account the results of the durability design and the required safety at the end of service life.

The combined method is especially suited to degradation mechanisms which directly affect the bearing capacity or the mechanical serviceability of structures. The method also, in some cases, allows use of smaller lifetime safety factors than in the separated method, as the value of $m$ is checked after the design. In safety class 2 (ultimate state) the required safety index after service life may be slightly smaller than in ordinary design.

The structures are provided with dimensions and material

specifications relevant to fulfilling the following condition:

$$R_d(t_d) - S_d(t_d) \geq 0 \qquad (6.12)$$

where

$R_d(t_d)$ = the design capacity of the structure at the end of the design service life, and

$S_d(t_d)$ = the design load of the structure at the end of the design service life.

The values for material and load safety factors depend on the safety class. In safety class 1 ($\beta = 3.8$) the material and load safety factors are the same as in ordinary design. In safety class 2 ($\beta = 3.1$) the following safety factors are used:

$$\gamma_g = 1.3$$
$$\gamma_p = 1.38$$
$$\gamma_c = 1.4$$
$$\gamma_s = 1.13$$

The final dimensions are readily obtained using spreadsheet applications with 'Goal seek' or 'Solver' tools.

After solution of the dimensions and material specifications for a structure the following control measures are performed. The condition for $m$ (relative reduction of the safety margin during $0 \rightarrow t_d$ is

$$m = \frac{\Theta_o - \Theta_m}{\Theta_o} \leq 0.7 \qquad (6.13)$$

In equation (6.13) $\Theta$ is the safety margin ($= R - S$) determined using characteristic values of loads and material properties. The index o refers to the initial state of the structure and $m$ to the final state after the design service life $t_d$. $\Theta_m$ is determined from the final mechanical design solution by setting the load and material safety factors equal to 1. $\Theta_o$ is obtained by also setting

$\gamma_t$ equal to 0. The calculations are easy when using a spreadsheet application (cf. examples in Section 6.3).

If $m$ is greater than 0.7 the required safety is probably not fulfilled, as the lifetime safety factor used in the calculations is too small for the real $m$ value. Also the durability designed structure is probably too robust compared with the respective ordinarily designed structure.

To keep $m$ smaller than required it may be necessary to return to phase 2 to make some modifications in the durability design. The designer may have to consider whether there are any means to reduce the rate of degradation or otherwise to change the design assumptions. If such is not possible, expectations of the service life of the structure may not be realistic in the intended environment and a reduced target service life may have to be considered.

If reduced values for material and load safety factors are used in the final design, a check must be performed that safety margin $\Theta_o$ and the bearing capacity $R_o$ (at the start of service life) are not smaller than those obtained by ordinary mechanical design. If they are, the dimensions obtained by ordinary mechanical design are determinative. The final safety of the structure must never be less than that of an ordinarily designed structure.

## 6.3 EXAMPLES OF THE DESIGN PROCEDURE

### 6.3.2 Column

**(a) Setting up the design problem**
The column is to be dimensioned for the following loads:

$$F_g = 1000 \text{ kN} \quad \text{(dead load)}$$
$$F_p = 3000 \text{ kN} \quad \text{(variable load)}$$

In a square cross-section (side length $b$) there are four steel bars ($\phi = D$), one in each corner. The yield strength of steel is 400 MPa ($= f_y$).

The column is supposed to be maintenance free so that

corrosion of steel bars in the assumed cracks or deterioration of the concrete cover do not hinder use of the column during its service life. The hoops (stirrups) must not be completely broken at cracks after the service life. The concrete cover must be at least 20 mm after the service life with no spalling due to general corrosion of hoops.

**(b) Ordinary mechanical design**
The ordinary design of the column is performed using the following formulae:

$$R_d \geq S_d \tag{6.14}$$

where

$$S_d = \gamma_g F_g + \gamma_p F_p \tag{6.15}$$

and

$$R_d = \frac{A_c f_c}{\gamma_c} + \frac{A_s f_y}{\gamma_s} \tag{6.16}$$

The cross-sectional areas of concrete and steel, $A_c$ and $A_s$, are

$$A_c = b^2 \tag{6.17}$$

$$A_s = \frac{4\pi D^2}{4} \tag{6.18}$$

Taking $D = 15$ mm we obtain from equation (6.18)

$$A_s = 707 \text{ mm}^2$$

As a result of the condition (6.14) we get

$$A_c = 210\,155 \text{ mm}^2$$

From equation (6.17) we get

$$b = 458 \text{ mm}$$

| Column | | | | | | |
|---|---|---|---|---|---|---|
| | Ordinary mechanical design | | | Durability and | | |
| | | Design values | Charact. values | | | |
| | | | | | final design | |
| | $\gamma_g$ | 1.35 | 1 | | $\gamma_t$ | 3.3 |
| | $\gamma_p$ | 1.5 | 1 | | $t_g$ | 50 |
| | $\gamma_c$ | 1.5 | 1 | | $t_d$ | 165 |
| | $\gamma_s$ | 1.15 | 1 | | $C_{min}$ | 20 |
| | $F_g$ | 1350000 | 1000000 | | $\delta L/\delta t$ | 0.135 |
| | $F_p$ | 4500000 | 3000000 | | $b'(t_d)$ | 22.3 |
| | $f_c$ | 27 | 40 | | $C_{omin}$ | 42 |
| | $f_y$ | 348 | 400 | | $C_a$(chosen) | 45 |
| | D | 15.0 | | | $b_{emin}$ | 503 |
| | $A_1$ | 707 | | | $b_a$(chosen) | 503 |
| | b | 458 | | | $D_{min}$ | 15 |
| | $A_c$ | 210155 | | | $\delta r/\delta t$ | 0.03 |
| | S | 5850000 | | | $d'(t_d)$ | 5.0 |
| | R | 5850000 | | | $D_{omin}$ | 24.9 |
| | R-S | -1.9E-09 | | | $D_a$(chosen) | 25 |
| | | | | | Check of the cracking | |
| | | | | | of concrete cover | |
| | | | | | C | 45 |
| | | | | | $D_{hmin}$ | 10 |
| | | | | | $D_h$(chosen) | 10 |
| | | | | | $C_h$ | 35 |
| | | | | | $c_{env}$ | 1 |
| | | | | | $c_a$ | 0.7 |
| | | | | | $f_{ck}$ | 40 |
| | | | | | a | 1800 |
| | | | | | b | -1.7 |
| | | | | | r | 12 |
| | | | | | $t_o$ | 401 |
| | | | | | $t_1$ | 30 |
| | | | | | $t_t$ | 431 |
| | | | | | $t_d$ | 165 |

**Figure 6.6** Spreadsheet design of a column. Separated design method.

These are the preliminary dimensions for steel bars and concrete.
   The calculations are also presented in Figure 6.6, which shows an excerpt from an Excel spreadsheet. The solution has been calculated by Solver by setting $R - S$ equal to 0 and changing the value of $b$.

## (c) Durability design

The target service life of a column is 50 years. The lifetime safety factor is 3.3 for the separated design method and 2.5 for the combined design service life. Thus the design service life is

$$t_d = 3.3 \times 50 = 165 \text{ years} \quad \text{(for the separated method)}$$
$$t_d = 2.5 \times 50 = 125 \text{ years} \quad \text{(for the combined method)}$$

The column is assumed to be partly immersed in river water that freezes in winter. Thus the depth of deterioration in concrete, $c'$, is evaluated using the following formula (cf. section 8.2):

$$c' = c_{env} c_{cur} c_{age} a^{-0.7} (f_{ck} + 8)^{-1.4} t \qquad (6.19)$$

The following values are chosen for the parameters in this formula:

$$c_{env} \text{ (environmental factor)} = 80$$
$$c_{cur} \text{ (curing factor)} = 1$$
$$c_{age} \text{ (ageing factor)} = 1$$
$$a \text{ (air content in \%)} = 4$$
$$f_{ck} \text{ (characteristic compressive strength MPa)} = 40$$

By inserting these values into equation (6.19) we get for the depth of deterioration of concrete

$$c' = 0.135t \text{ (mm)} \qquad (6.20)$$

The depth of corrosion in steel bars at cracks is evaluated as follows (cf. section 8.5):

$$d' = 0.03t \text{ (mm)} \qquad (6.21)$$

Depending on the design service life we get the following durability design parameters.

*Separated design method*
($t_d$ = 165 years)
The depth of deterioration of concrete is

$$c' = 0.135 \times 165 = 22.3 \text{ mm}$$

The minimum thickness of concrete cover is

$$C_{min} = 20 + 22.3 = 43 \text{ mm}$$

We choose $C = 45$ mm.
   The depth of corrosion of steel bars at cracks is

$$d' = 0.03 \times 165 = 5.0 \text{ mm}$$

The hoops must not be broken due to corrosion. Thus the minimum diameter of the hoops is

$$D_{hmin} = 2 \times 5.0 \text{ mm} = 10 \text{ mm}$$

We choose $D_h = 10$ mm.

*Combined design method*
($t_d$ = 125 years)

$$c' = 0.135 \times 125 = 16.9 \text{ mm}$$

$$C_{min} = 20 + 16.9 = 37 \text{ mm}$$

We choose $C = 40$ mm.

$$d' = 0.03 \times 125 = 3.8 \text{ mm}$$

$$D_{hmin} = 2 \times 3.8 \text{ mm} = 8 \text{ mm}$$

We choose $D_h = 10$ mm.

A check for general corrosion with the chosen concrete cover and diameter of steel bars must still be performed. The chloride content of water is assumed to be so low that corrosion cannot be induced by chlorides. However, steel bars in the upper end of the columns may be corroded due to carbonation.

The determinative reinforcement with respect to general corrosion is the hoops. Thus equations (8.20) and (8.29) (cf. section 8.5) are applied to hoops (the diameter of the hoops is subtracted from the concrete cover of the main reinforcement):

$$\mu(t_L) = \frac{C_h^2}{\left(c_{env}c_{air}a(f_{ck}+8)^b\right)^2} + \frac{80C_h}{D_h r} \tag{6.22}$$

where

$\quad C_h =$ the concrete cover with respect to hoops ($= C - D_h$),

$\quad D_h =$ the diameter of hoops,

$\quad C_{env} =$ the environmental coefficient ($= 1$),

$\quad C_{air} =$ the coefficient of air content ($= 0.7$), and

$\quad\quad r =$ the rate of corrosion of steel before cracking

$\quad\quad\quad$ ($\sim 12\ \mu m/year$).

As a result we get $\mu(t_L) = 431$ years for the separated design method, which is longer than the design service life (165 years). For the combined design method we get 322 years which is also longer than the design service life (125 years). Therefore the chosen concrete covers are acceptable.

## (d) Final design

*Separated design method*
The width of the column at the start of service life is determined by adding twice the depth of deterioration of concrete to the width obtained in the ordinary design (phase 1):

$$b_o = b + 2c' = 458\ \text{mm} + 2 \times 22.3\ \text{mm} = 503\ \text{mm}$$

The diameter of steel bars of the main reinforcement is determined in the same way:

$$D_{omin} = 15\,\text{mm} + 2 \times 5.0\,\text{mm} = 25.0\,\text{mm}$$

As the diameter matches the standard we choose 25 mm. The calculations are presented in Figure 6.6.

*Note:* The required thickness of the concrete cover ($C_{min}$) and the width of the column ($b_o$) could be reduced by increasing the compressive strength of the concrete. For $f_{ck} = 50\,\text{MPa}$ the minimum thickness of concrete cover would be 37 mm and the width of the column 444 mm.

*Combined design method*
When the combined method is used, dimensioning of the column is performed by applying the ordinary design formulae at $t = t_d$:

$$R_d(t_d) \geq S_d \tag{6.23}$$

where

$$S_d = \gamma_g F_g + \gamma_p F_p \tag{6.24}$$

and

$$R_d(t_d) = A_c(t_d)\frac{f_c}{\gamma_c} + A_s(t_d)\frac{f_y}{\gamma_s} \tag{6.25}$$

The cross-sectional areas of concrete and steel, $A_c$ and $A_s$, are

$$A_c(t_d) = (b_o - 2c'(t_d))^2 \tag{6.26}$$

$$A_s(t_d) = \frac{4\pi(D_o - 2d'(t_d))^2}{4} \tag{6.27}$$

For $D_o = 25\,\text{mm}$ we get from equation (6.27)

$$A_s(t_d) = 962\,\text{mm}^2$$

| Column | | | | | | | |
|---|---|---|---|---|---|---|---|
| **Ordinary mechanical design** | | | | **Durability and final design** | | | |
| | Design values | Charact. values | | | Design values | Charact. values | Charact. values |
| | $t=0$ | $t=0$ | | | $t=t_d$ | $t=t_d$ | $t=0$ |
| $\gamma_a$ | 1.35 | 1 | | $\gamma_a$ | 1.35 | 1 | 1 |
| $\gamma_p$ | 1.5 | 1 | | $\gamma_p$ | 1.5 | 1 | 1 |
| $\gamma_c$ | 1.5 | 1 | | $\gamma_c$ | 1.5 | 1 | 1 |
| $\gamma_v$ | 1.15 | 1 | | $\gamma_v$ | 1.15 | 1 | 1 |
| $F_a$ | 1350000 | 1000000 | | $\gamma_t$ | 2.5 | 2.5 | 0 |
| $F_p$ | 4500000 | 3000000 | | $F_a$ | 1350000 | 1000000 | 1000000 |
| $f_c$ | 27 | 40 | | $F_p$ | 4500000 | 3000000 | 3000000 |
| $f_v$ | 348 | 400 | | $f_c$ | 27 | 40 | 40 |
| $D$ | 15.0 | 15.0 | | $f_v$ | 348 | 400 | 400 |
| $A_s$ | 707 | 707 | | $C_{mn}$ | 20 | 20 | 20 |
| $b$ | 458 | 458 | | $D_{mn}$ | 15 | 15 | 15 |
| $A_c$ | 210155 | 210155 | | $t_a$ | 50 | 50 | 50 |
| $S$ | 5850000 | 4000000 | | $t_d$ | 125 | 125 | 0 |
| $R$ | 5850000 | 8688948 | | $\delta L/\delta t$ | 0.135 | 0.135 | 0.135 |
| $R-S$ | 0.1322364 | 4688948 | | $b'(t_d)$ | 16.9 | 16.9 | 0.0 |
| | | | | $C_{omn}$ | 37 | 37 | 20 |
| | | | | $C_o$(chosen) | 40 | 40 | 40 |
| **Check of the cracking** | | | | $\delta r/\delta t$ | 0.03 | 0.03 | 0.03 |
| **of concrete cover** | | | | $d'(t_d)$ | 3.8 | 3.8 | 0.0 |
| $C$ | 40 | | | $D_{omn}$ | 22.5 | 22.5 | 15.0 |
| $D_{hvmin}$ | 8 | | | $D_o$(chosen) | 25 | 25 | 25 |
| $D_h$(chosen) | 10 | | | $D(t_d)$ | 17.5 | 17.5 | 25.0 |
| $C_h$ | 30 | | | $A_s(t_d)$ | 962 | 962 | 1963 |
| $C_{env}$ | 1 | | | $b_o$ | 489 | 489 | 489 |
| $C_{qr}$ | 0.7 | | | $A_c(t_d)$ | 206826 | 206826 | 238662 |
| $f_{ck}$ | 40 | | | $S$ | 5850000 | 4000000 | 4000000 |
| $a$ | 1800 | | | $R$ | 5850000 | 8657873 | 10331897 |
| $b$ | -1.7 | | | $R-S$ | -0.043762 | 4657873 | 6331897 |
| $r$ | 12 | | | | | | |
| $t_o$ | 295 | | | **Check of structural degradation** | | | |
| $t_1$ | 27 | | | Rel. reduction of R | 0.162 | | |
| $t_L$ | 322 | | | Rel. reduction of R-S | 0.264 | | |
| $t_d$ | 125 | | | | | | |

**Figure 6.7** Spreadsheet design of a column. Combined design method.

As a result of the condition from equation (6.23) we get

$$A_c(t_d) = 206\,826 \text{ mm}^2$$

From equation (6.26) we obtain

$$(b_o - 2 \times 16.9)^2 = 206\,826 \text{ mm}^2$$

which yields

$$b_o = 489 \text{ mm}$$

The calculations are presented in Figure 6.7. The solution has been sought by Solver by setting $R - S$ equal to 0 and changing the value of $b_o$. In the next column the safety margin $\Theta_m = R_k - S_k$ corresponding to the solution obtained has been determined by inserting $\gamma_g = \gamma_p = \gamma_c = \gamma_s = 1$:

$$\Theta_m = 4658 \text{ kN}$$

By also inserting $\gamma_t = 0$ we get $\Theta_o$ representing the safety margin at the start of service life $(t = 0)$:

$$\Theta_o = 6332 \text{ kN}$$

The relative reduction of $R - S(= m)$ is

$$m = (\Theta_o - \Theta_m)/\Theta_o = 0.264$$

which is smaller than 0.7.

The width of the column obtained by the combined method is a little smaller than that obtained by the separated design method.

### 6.3.3 Beam

#### (a) Setting up the design problem
The beam is to be dimensioned for the following loads:

$$M_g = 10 + 0.1d \text{ kNm} \quad (d \text{ in mm})$$
$$M_p = 50 \text{ kNm}$$

The cross-section of the beam is assumed to be rectangular with width $b$ ($\approx 300$ mm) and efficient height $d$. At the lower edge of the beam are three steel bars. The yield strength of steel is 400 MPa. The characteristic compressive strength is 40 MPa, the air content is 2% (not air-entrained), and the binding agent is Portland cement.

The beam is supposed to be maintenance free so that corrosion of steel bars in the assumed cracks or degradation of the concrete cover will not hinder use of the column during its service life. The cross-section of hoops (stirrups) must not be fully corroded at cracks. The concrete cover must be at least 20 mm after the service life with no spalling due to general corrosion.

#### (b) Ordinary mechanical design
The ordinary mechanical design of the beam is performed using traditional design principles:

$$R_d \geq S_d \tag{6.28}$$

$$S_d = \gamma_g M_g + \gamma_p M_p \tag{6.29}$$

$$R_{ds} = A_s Z \frac{f_y}{\gamma_s} \quad \text{(the stress of steel is decisive)} \tag{6.30}$$

$$R_{dc} = bxz \frac{f_c}{2\gamma_c} \quad \text{(the stress of concrete is decisive)} \tag{6.31}$$

$$x = d\mu n \left[ -1 + \left( 1 + \frac{2}{\mu n} \right)^{1/2} \right] \tag{6.32}$$

$$z = d - 0.4x \tag{6.33}$$

$$n = \frac{E_s}{E_c}$$

$$\mu = \frac{A_s}{bd} \tag{6.34}$$

$A_s$ is the cross-sectional area of steel bars:

$$A_s = 3\pi \frac{D^2}{4} \tag{6.35}$$

Taking $D = 15$ mm we get

$$A_s = 530 \text{ mm}^2$$

By setting $R_{ds}$ equal to $S_d$ we get

$$d = 2083 \text{ mm}$$

However, increasing the diameter of the steel bars quickly reduces the efficient height. By changing $D$ to 20 mm we get

$$A_s = 942 \text{ mm}^2$$
$$d = 543 \text{ mm}$$

The calculations are also seen in Figure 6.8.

## (c) Durability design
The target service life is 50 years. The lifetime safety factor is 3.3 for the separated design method and 2.5 for the combined design method. Thus the design service life, $t_d$, is 165 years for the separated method and 125 years for the combined method.

| Beam | | | | | | |
|------|--------|--------|---|------|------|---|
| | Ordinary mechanical design | | | Durability and | | |
| | Design values | Charact. values | | final design | | |
| | | | | | | |
| $\gamma_g$ | 1.35 | 1 | | $\gamma_t$ | 3.3 | |
| $\gamma_p$ | 1.5 | 1 | | $t_o$ | 50 | |
| $\gamma_c$ | 1.5 | 1 | | $t_d$ | 165 | |
| $\gamma_s$ | 1.15 | 1 | | $C_{min}$ | 20 | |
| $M_g$ | 86813146 | 64306034 | | $\delta L/\delta t$ | 0.117 | |
| $M_p$ | 75000000 | 50000000 | | $d'(t_d)$ | 19.3 | |
| $f_c$ | 27 | 40 | | $b'(t_d)$ | 19.3 | |
| $f_y$ | 348 | 400 | | $C_{omin}$ | 39.3 | |
| D | 20.0 | | | $C_o$(chosen) | 40 | |
| $A_s$ | 942 | | | $b_{omin}$ | 339 | |
| b | 300 | | | $b_o$(chosen) | 339 | |
| d | 543 | | | $d_{omin}$ | 562 | |
| n | 5.80 | | | $d_o$(chosen) | 562 | |
| m | 0.0058 | | | $D_{min}$ | 20.0 | |
| x | 124 | | | $\delta r/\delta t$ | 0.03 | |
| z | 494 | | | $d'(t_d)$ | 5.0 | |
| S | 161813146 | | | $D_{omin}$ | 29.9 | |
| $R_c$ | 244109121 | | | $D_o$(chosen) | 30 | |
| $R_s$ | 161813146 | | | | | |
| R | 161813146 | | | | | |
| R-S | 2.98E-08 | | | Check of the cracking | | |
| | | | | of concrete cover | | |
| | | | | C | 40 | |
| | | | | $D_{hmin}$ | 10 | |
| | | | | $D_h$(chosen) | 10 | |
| | | | | $C_h$ | 30 | |
| | | | | $c_{env}$ | 1 | |
| | | | | $c_{qt}$ | 1 | |
| | | | | $f_{ck}$ | 40 | |
| | | | | a | 1800 | |
| | | | | b | -1.7 | |
| | | | | r | 12 | |
| | | | | $t_o$ | 145 | |
| | | | | $t_1$ | 20 | |
| | | | | $t_k$ | 165 | |
| | | | | $t_d$ | 165 | |

**Figure 6.8** Spreadsheet design of a beam. Separated design method.

All sides of the beam are assumed to be exposed to frost action. The environmental factor for frost attack, $c_{env}$, is 40 and the anticipated curing time is 3 days.

The curing factor is (cf. equation (8.4), section 8.2.2)

$$c_{cur} = \frac{1}{0.85 + 0.17 \log_{10}(3)} = 1.074 \qquad (6.36)$$

As the concrete is made of Portland cement we conclude:

$$c_{age} = 1$$

Inserting these values into equation (6.19) we get

$$c' = 0.117t \text{ (mm)} \qquad (6.37)$$

At the same time corrosion is occurring in steel bars at cracks. The rate of corrosion is evaluated as 0.03 mm/year:

$$d' = 0.03t \text{ (mm)} \qquad (6.38)$$

The durability design parameters are as follows (depending on the design service life).

*Separated design method*
($t_d$ = 165 years)
The depth of deterioration is

$$c' = 0.117 \times 165 = 19.3 \text{ mm}$$

The required concrete cover is

$$C_{min} = 20 + 19.3 \text{ mm} = 39.3 \text{ mm}$$

We choose $C$ = 40 mm.
The depth of corrosion at cracks is

$$d' = 0.03 \times 165 = 5.0 \text{ mm}$$

The diameter of hoops must be at least

$$D_{hmin} = 2 \times 5.0 \, mm = 10.0 \, mm$$

We choose $D_h = 10$ mm.

*Combined method*
($t_d = 125$ years)

$$c' = 0.117 \times 125 = 14.6 \, mm$$
$$C_{min} = 20 + 14.6 \, mm = 34.6 \, mm$$

We choose $C = 35$ mm.

$$d' = 0.03 \times 125 = 3.75 \, mm$$
$$D_{hmin} = 2 \times 3.75 \, mm = 7.5 \, mm$$

We choose $D_h = 10$ mm.

The cracking time of the concrete cover is then checked by applying equation (6.22). The following values of parameters are inserted into the formula:

$$C = 40 \, mm \quad \text{(separated design method)}$$
$$\text{or 35 (combined design method)}$$
$$C_h = C - D_h = C - 10 \, mm$$
$$f_{ck} = 40 \, MPa$$
$$c_{env} = 1$$
$$c_{air} = 1$$
$$D_h = 10 \, mm$$
$$r = 12 \, \mu m$$

With the separated method we get $\mu(t_L) = 165$ years, which equals the design service life (165 years). Thus the concrete cover of 40 mm is adequate. With the combined method we get $\mu(t_c) = 117$ years, which is less than the design service life (125 years).

Therefore the concrete cover is increased from 35 to 40 mm. Then the calculated service life is 165 years, which fulfils the requirement.

## (d) Final design

*Separated design method*
The width of the beam at the start of service life is twice the deterioration depth of concrete added to the width obtained in ordinary design:

$$b_o = b + 2b' = 300 + 2 \times 19.3 = 339 \text{ mm}$$

The effective height of the beam is increased by the depth of deterioration:

$$d_o = d + b' = 543 + 19.3 \text{ mm} = 562 \text{ mm}$$

The minimum diameter of the steel bars is

$$D_{omin} = 20 + 2 \times 5.0 \text{ mm} = 30.0 \text{ mm}$$

We choose $D_o = 30$ mm.
The calculations are presented in Figure 6.8.

*Combined design method*
Dimensioning is performed using the following formulae:

$$R_d(t_d) \geq S_d \qquad\qquad (6.39)$$

where

$$S_d = \gamma_g M_g + \gamma_p M_p \qquad\qquad (6.40)$$

and $R_d$ is the smaller of the following quantities:

$$R_{ds} = A_s(t_d) z(t_d) \frac{f_y}{\gamma_s} \quad \text{(the stress of steel is decisive)} \qquad (6.41)$$

$$R_{dc} = b(t_d)x(t_d)z(t_d)\frac{f_c}{2\gamma_c} \quad \text{(the stress of concrete is decisive)} \quad (6.42)$$

where

$$x(t_d) = d(t_d)\mu(t_d)n\left[-1 + \left(1 + \frac{2}{\mu(t_d)n}\right)^{1/2}\right] \quad (6.43)$$

$$z(t_d) = d(t_d) - 0.4x(t_d) \quad (6.44)$$

$$n = \frac{E_s}{E_c}$$

$$\mu(t_d) = \frac{A_s(t_d)}{b(t_d)d(t_d)} = \frac{N_s\pi(D_o - d'(t_d))^2/4}{(b_o - 2c'(t_d))(d_o - c'(t_d))} \quad (6.45)$$

$A_s$ is the cross-sectional area of steel:

$$A_s(t_d) = \frac{3\pi(D_o - 2d'(t_d))^2}{4} \quad (6.46)$$

With $D_o = 28$ mm and $d'(t_d) = 3.75$ mm we get

$$A_s(t_d) = 990 \text{ mm}^2$$

The effective height of the beam, $d_o$, is now dimensioned by setting $R_{ds}$ or $R_{dc}$, whichever is smaller, equal to $S_d$ at $t = t_d$. This gives (cf. Figure 6.9)

$$R_{ds} = S_d => d_o = 531 \text{ mm}$$

The width of the beam is 300 mm and the concrete cover is 40 mm with respect to the main reinforcement.

The effective height is slightly smaller than that determined by the separated design method. The relative reduction of safety margin ($m = 0.693$) fulfils the maximum requirement of 0.7.


*Combined design method in safety class 2*
In safety class 2 (ultimate state) the structural durability design is

| Beam | | | | | | | |
|---|---|---|---|---|---|---|---|
| Ordinary mechanical design | | | | Durability and final design | | | |
| | Design values | Charact. values | | | Design values | Charact. values | Charact. values |
| | $t=0$ | $t=0$ | | | $t=t_d$ | $t=t_d$ | $t=0$ |
| $\gamma_g$ | 1.35 | 1 | | $\gamma_g$ | 1.35 | 1 | 1 |
| $\gamma_p$ | 1.5 | 1 | | $\gamma_p$ | 1.5 | 1 | 1 |
| $\gamma_c$ | 1.5 | 1 | | $\gamma_c$ | 1.5 | 1 | 1 |
| $\gamma_s$ | 1.15 | 1 | | $\gamma_s$ | 1.15 | 1 | 1 |
| | | | | $\gamma_L$ | 2.5 | 2.5 | 0 |
| $M_g$ | 86813137 | 64306027 | | $M_g$ | 85195765 | 63107974 | 63107974 |
| $M_p$ | 75000000 | 50000000 | | $M_p$ | 75000000 | 50000000 | 50000000 |
| $f_c$ | 27 | 40 | | $f_c$ | 27 | 40 | 40 |
| $f_y$ | 348 | 400 | | $f_y$ | 348 | 400 | 400 |
| D | 20.0 | 20.0 | | $C_{min}$ | 20 | 20 | 20 |
| $A_s$ | 942 | 942 | | $D_{min}$ | 20 | 20 | 20 |
| b | 300 | 300 | | $t_a$ | 50 | 50 | 50 |
| d | 543 | 543 | | $t_d$ | 125 | 125 | 0 |
| n | 5.80 | 5.80 | | $\delta L/\delta t$ | 0.117 | 0.117 | 0.117 |
| m | 0.0058 | 0.0058 | | $d'(t_a)$ | 14.6 | 14.6 | 0.0 |
| x | 124 | 124 | | $b'(t_a)$ | 14.6 | 14.6 | 0.0 |
| z | 494 | 494 | | $C_{emin}$ | 34.6 | 34.6 | 20.0 |
| S | 161813137 | 114306027 | | $C_e$(chosen) | 40 | 40 | 40 |
| $R_c$ | 244109073 | 366163609 | | $\delta r/\delta t$ | 0.03 | 0.03 | 0.03 |
| $R_s$ | 161813125 | 186085094 | | $d'(t_a)$ | 3.8 | 3.8 | 0.0 |
| R | 161813125 | 186085094 | | $D_{emin}$ | 27.5 | 27.5 | 20.0 |
| R-S | -11.76266 | 71779067 | | $D_e$(chosen) | 28 | 28 | 28 |
| | | | | $D(t_a)$ | 20.5 | 20.5 | 28.0 |
| | | | | $A_s(t_a)$ | 990 | 990 | 1847 |
| | | | | $b_e$ | 300 | 300 | 300 |
| Check of the cracking | | | | $d_e$ | 531 | 531 | 531 |
| of concrete cover | | | | $b(t_a)$ | 271 | 271 | 300 |
| C | 40 | | | $d(t_a)$ | 516 | 516 | 531 |
| $D_{hmin}$ | 8 | | | n | 5.80 | 5.80 | 5.80 |
| $D_h$(chosen) | 10 | | | $m(t_a)$ | 0.0071 | 0.0071 | 0.0116 |
| $C_h$ | 30 | | | $x(t_a)$ | 128 | 128 | 162 |
| $C_{env}$ | 1 | | | $z(t_a)$ | 465 | 465 | 466 |
| $C_{at}$ | 1 | | | S | 160195765 | 113107974 | 113107974 |
| $f_{ck}$ | 40 | | | $R_c$ | 215468243 | 323202365 | 453953369 |
| a | 1800 | | | $R_s$ | 160195764 | 184225129 | 344445782 |
| b | -1.7 | | | R | 160195764 | 184225129 | 344445782 |
| r | 12 | | | R-S | -1.0721515 | 71117155 | 231337807 |
| $t_o$ | 145 | | | | | | |
| $t_1$ | 20 | | | Check of structural degradation | | | |
| $t_L$ | 165 | | | Relative reduction of R | | | 0.465 |
| $t_d$ | 125 | | | Relative reduction of R-S | | | 0.693 |

**Figure 6.9** Spreadsheet calculations for a beam. Combined design method.

carried out by applying the value 2.2 for $\gamma^t$ and reduced values for material and load safety factors (cf. section 5.3.4). The calculations are presented in Figure 6.10. The effective height of the beam is 414 mm which is clearly smaller than in safety class 1 (531 mm); 35 mm is adequate for the cover. The relative reduction of $\Theta$ is 0.681, which is smaller than 0.7.

When reduced values of load and material safety factors are used it is essential to check also that the structural capacity and the safety margin determined by the final design are not smaller than those of the ordinary design.

Durability design:

$$R_m = 265 \text{ kNm}$$

and

$$\Theta_m = 163 \text{ kNm}$$

Ordinary design:

$$R_o = 186 \text{ kNm}$$

and

$$\Theta_o = 72 \text{ kNm}$$

In this case the values of $R_m$ and $\Theta_m$ of the durability design are both greater than those of the ordinary design. Thus the dimensions of the final (durability) design are determinative.

| Beam | | | | | | | |
|---|---|---|---|---|---|---|---|
| Ordinary mechanical design | | | | Durability and final design | | | |
| | Design values | Charact. values | | | Design values | Charact. values | Charact. values |
| | $t=0$ | $t=0$ | | | $t=t_d$ | $t=t_d$ | $t=0$ |
| $\gamma_g$ | 1.35 | 1 | | $\gamma_g$ | 1.3 | 1 | 1 |
| $\gamma_p$ | 1.5 | 1 | | $\gamma_p$ | 1.38 | 1 | 1 |
| $\gamma_c$ | 1.5 | 1 | | $\gamma_c$ | 1.4 | 1 | 1 |
| $\gamma_s$ | 1.15 | 1 | | $\gamma_s$ | 1.13 | 1 | 1 |
| | | | | $\gamma_t$ | 2.2 | 2.2 | 0 |
| $M_g$ | 86813137 | 64306027 | | $M_g$ | 66866942 | 51436110 | 51436110 |
| $M_p$ | 75000000 | 50000000 | | $M_p$ | 69000000 | 50000000 | 50000000 |
| $f_c$ | 27 | 40 | | $f_c$ | 29 | 40 | 40 |
| $f_y$ | 348 | 400 | | $f_y$ | 354 | 400 | 400 |
| D | 20.0 | 20.0 | | $C_{min}$ | 20 | 20 | 20 |
| $A_s$ | 942 | 942 | | $D_{min}$ | 20 | 20 | 20 |
| b | 300 | 300 | | $t_a$ | 50 | 50 | 50 |
| d | 543 | 543 | | $t_d$ | 110 | 110 | 0 |
| n | 5.80 | 5.80 | | $\delta L/\delta t$ | 0.117 | 0.117 | 0.117 |
| m | 0.0058 | 0.0058 | | $d'(t_d)$ | 12.9 | 12.9 | 0.0 |
| x | 124 | 124 | | $b'(t_d)$ | 12.9 | 12.9 | 0.0 |
| z | 494 | 494 | | $C_{omh}$ | 32.9 | 32.9 | 20.0 |
| S | 161813137 | 114306027 | | $C_a$(chosen) | 35 | 35 | 35 |
| $R_c$ | 244109073 | 366163609 | | $\delta r/\delta t$ | 0.03 | 0.03 | 0.03 |
| $R_s$ | 161813125 | 186085094 | | $d'(t_d)$ | 3.3 | 3.3 | 0.0 |
| R | 161813125 | 186085094 | | $D_{amh}$ | 26.6 | 26.6 | 20.0 |
| R-S | -11.76266 | 71779067 | | $D_a$(chosen) | 28 | 28 | 28 |
| | | | | $D(t_d)$ | 21.4 | 21.4 | 28.0 |
| | | | | $A_s(t_d)$ | 1079 | 1079 | 1847 |
| | | | | $b_o$ | 300 | 300 | 300 |
| Check of the cracking | | | | $d_o$ | 414 | 414 | 414 |
| of concrete cover | | | | $b(t_d)$ | 274 | 274 | 300 |
| C | 35 | | | $d(t_d)$ | 401 | 401 | 414 |
| $D_{hmin}$ | 7 | | | n | 5.80 | 5.80 | 5.80 |
| $D_h$(chosen) | 10 | | | $m(t_d)$ | 0.0098 | 0.0098 | 0.0149 |
| $C_h$ | 25 | | | $x(t_d)$ | 114 | 114 | 140 |
| $C_{env}$ | 1 | | | $z(t_d)$ | 356 | 356 | 358 |
| $C_{at}$ | 1 | | | S | 135866942 | 101436110 | 101436110 |
| $f_{ck}$ | 40 | | | $R_c$ | 159515798 | 223322118 | 301012942 |
| a | 1800 | | | $R_s$ | 135866939 | 153529642 | 264795633 |
| b | -1.7 | | | R | 135866939 | 153529642 | 264795633 |
| r | 12 | | | R-S | -2.9321909 | 52093532 | 163359523 |
| $t_o$ | 100 | | | | | | |
| $t_1$ | 17 | | | Check of structural degradation | | | |
| $t_L$ | 117 | | | Relative reduction of R | | | 0.420 |
| $t_d$ | 110 | | | Relative reduction of R-S | | | 0.681 |

**Figure 6.10** Spreadsheet calculations for a beam designed in safety class 2 by the combined design method.

# 7

# Durability models

## 7.1 TYPES OF DURABILITY MODELS

### 7.1.1 Degradation, performance and service life models

For structural durability design a designer needs durability models with which he can evaluate the time-related changes in materials and structures. These models include design parameters such as structural dimensions, material properties, environmental parameters, etc. The durability model concept covers different types of models that are specified more closely in the following.

Mathematical presentations that show an increase in degradation with time (or age) and with appropriate design parameters are called degradation models. They are used in durability design when the limit state is expressed as maximum degradation.

Degradation can alternatively be presented as a decrease in performance. Mathematical presentations that show decreased performance as a function of time and appropriate design parameters are called performance models. Performance models are used in durability design when the limit state is expressed as the minimum performance.

Mathematical presentations that show the service life of a structure as a function of different design parameters are called service life models. They can often be derived from degradation

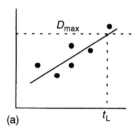

(a)

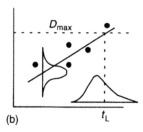

(b)

**Figure 7.1**    (a) Deterministic and (b) stochastic models.

or performance models when the limit states of maximum degradation or minimum performance are known. Service life models are used whenever the design problem can be formulated according to the service life principle (cf. section 3.3).

There may be durability models for different levels such as materials, structural elements or buildings. All of these can be used in durability design. Structural level models such as those for bearing capacity are used in structural design. They are usually created by incorporating degradation models of materials in the basic design formulae of structural design.

### 7.1.2 Deterministic and stochastic durability models

Durability models can also be divided into deterministic or stochastic models (Figure 7.1). Deterministic durability models are used in deterministic durability design where the scatter of degradation (or performance or service life) is not taken into account. With known values of parameters the models yield only one value (of degradation or performance or service life) which is often the mean value. In some cases, deterministic models are formulated to give an upper or lower fractile value instead of the mean.

In many cases the information yielded by deterministic models is insufficient to evaluate the risk of not reaching the target service life. Especially in the mechanical design of structures, stochastic design methods are considered essential as the scatter due to degradation is normally wide and the degree of risk may be great.

When using stochastic durability models, the structures are designed by ensuring a certain minimum reliability with respect to target service life. The target service life is not an absolute value which must be met at all costs. Rather the requirement is that the probability of the service life falling short of the target service life is smaller than the allowed failure probability. In this way the real nature of service life is better considered and the design corresponds more closely to the expected result.

Stochastic design methods also provide a designer with the possibility of evaluating the sensitivity of different parameters affecting service life. Thus the main attention can be directed to these parameters. This possibility may be valuable even at the developmental stage of modelling.

## 7.1.3 Durability models for different purposes

### (a) Identification of needs for modelling

Durability models are produced for different purposes. The premises and aims for creating durability models are different depending on the purpose. There are various types and levels of durability data on which the models are based. Sometimes the prediction of service life can be based on the history of, and experience with, the structures, while in other cases tests must be performed to obtain necessary data for such predictions.

There are also differences in the parameters of models depending on the theoretical backgrounds of the models and the requirements of the user. For instance, parameters for design and quality control are different from those for surveillance and repair.

At least the following needs for durability models can be identified:

1. technical material development;
2. ecological evaluation of materials;
3. network level management systems for the maintenance, repair and rehabilitation of structures;
4. planning of project level repairs;
5. risk analysis of important structures;
6. design of a material mix and quality assurance at the construction site;
7. structural durability design.

There is no expectation concerning the weathering resistance of new materials. To prove the resistance of new materials, accelerated ageing tests are performed. To evaluate the service life of a new material an assumption is made that the number of cycles in an accelerated ageing test bears some kind of relationship to the lifetime of that material in actual conditions. By comparing the rate of change of material performance in an accelerated ageing test with that observed in a long-term ageing test under in-use conditions, the service life of a new material can be evaluated (Masters and Brandt, 1989). Comparative tests with some known materials may also serve as a basis for service life prediction (Pihlajavaara, 1984).

For ecological evaluation of materials the prediction of service life may also be based on accelerated ageing tests. The difference is, however, that the range of prediction may be much longer. For example, for the final repository conditions of nuclear waste the required service life of materials may be thousands of years.

Durability models are also needed for planning the maintenance policies and strategies of existing structures with cost analysis. Network level management systems are created for purposes of maintenance, repair and rehabilitation of bridges and other structures. Durability models for such systems are often based on inspection data collected by the agency for which the system is created. The parameters of durability models are then dependent on the data items with which the inspection data can be differentiated in the data base system.

Other kinds of models may be used in planning the repair work on existing single structures. By knowing the age and present condition of the structure, the prediction of future performance and remaining service life is usually possible using an extrapolation model.

Risk analyses are performed for structures associated with great social, economic and ecological risks. Such structures include nuclear power plants, oil platforms, dams, bridges, etc. For risk analysis it is essential to combine different plausible failure modes to determine the critical paths that would violate the security of the structure. All kinds of models, even theoretical, can be used in risk analysis.

Durability models are also needed in the design of concrete mixes. The parameters of such models would be the amounts and proportions of mix ingredients and properties that can be measured from a fresh mix. These models can also be used at the construction site. With the aid of durability models a supervisor of concrete construction work can evaluate the quality of concrete before allowing it to be placed in moulds.

**(b) Models for structural durability design**
Parameters of models to be used in structural durability design must be appropriate for a structural designer. Instead of parameters relating to a fresh concrete mix, parameters that are measurable from hardened concrete such as strength, porosity, etc. are preferred. They should also be included in the quality control system of the designed structures.

The strength properties of materials strongly influence the load-bearing capacity of structures and indirectly also the dimensions, span lengths, deformations, etc. As the strength properties already belong to the design parameters of structural design they are also very suitable as parameters of durability models.

If strength properties are not used in the durability models of concrete structures, it is highly probable that other parameters

**Table 7.1** Degradation factors and processes

| Degradation factor | Process | Degradation |
|---|---|---|
| **Mechanical** | | |
| Static loading | Deformation | Deflection, cracking, failure |
| Cyclic loading | Fatigue, deformation | Deflection, cracking, failure |
| Impact loading | Fatigue | Vibration, deflection, cracking, failure |
| **Biological** | | |
| Micro-organisms | Acid production | Leaching |
| **Chemical** | | |
| Soft water | Leaching | Disintegration of concrete |
| Acid | Leaching | Disintegration of concrete |
| Acid | Neutralization | Steel depassivation[a] |
| Acidifying gases | Neutralization | Steel depassivation[a] |
| Carbon dioxide | Carbonation | |
| Sulphur dioxide | | |
| Nitrogen dioxide | | |
| Chlorides | Penetration, destruction of passive film | Steel depassivation[a] |
| Steel depassivation, oxygen, water | Corrosion | Expansion of steel, loss of diameter in rebars, loss of bond |
| Stress/chlorides | Stress corrosion | Failure in prestressing tendons |
| Sulphates | Crystal pressure | Disintegration of concrete |
| Silicate aggregate, alkalis | Silicate reaction | Expansion, disintegration |
| Carbonate aggregate | Carbonate reaction | Expansion, disintegration |

**Table 7.1**  Degradation factors and processes (continued)

| Degradation factor | Process | Degradation |
|---|---|---|
| **Physical** | | |
| Temperature change | Expansion | Shortening, lengthening, restricted deformation |
| RH change | Shrinkage, swelling | Shortening, lengthening, restricted deformation |
| Low temperature, water | Ice formation | Disintegration of concrete |
| Deicing salt, frost | Heat transfer | Scaling of concrete |
| Floating ice | Abrasion | Cracking, scaling |
| Traffic | Abrasion | Rutting, wearing, tearing |
| Running water | Erosion | Surface damage |
| Turbulent water | Cavitation | Caves |

[a] Indicates 'intermediate state'.

with a relationship to strength parameters are used instead. This leads to a risk of contradictory requirements of design parameters. For this reason, the water–cement ratio of concrete, which is often used as a design parameter of concrete proportioning, is not considered suitable for structural design. Another reason for rejecting it as a parameter of structural design is that cement replacements such as blast furnace slag, fly ash and silica fume have been increasingly incorporated into concrete mixes, obscuring the whole concept of the water–cement ratio. Also, the control of the water–cement ratio for hardened concrete is difficult especially if cement replacements are used.

The degree of hydration of cement is another parameter often used for characterizing the quality of concrete (together with the water–cement ratio). Because of cement replacements the concept of degree of hydration has also become obscured, its value as a practical design parameter being reduced. For a structural designer the length of curing time is more meaningful and precise enough for design purposes.

## 7.2 DEVELOPMENT OF DURABILITY MODELS FOR DURABILITY DESIGN

### 7.2.1 Qualitative analysis of degradation

The first step in the process of producing durability models is the analysis of degradation factors and processes. All possible degradation factors, processes and effects are listed systematically in a table, as exemplified in Table 7.1. For convenience the degradation factors are subdivided into:

1. mechanical
2. biological
3. chemical
4. physical
5. use.

It is not always possible in this way to make clear distinctions between degradation factors. An example of this problem is biogenic sulphur attack, in which the origin is biological but the degradation is chemical. In such cases the type of degradation factor is based on the type of origin. The main purpose of classifying degradation factors is to present them in a clear overview.

A degradation table, such as Table 7.1 (originally used for risk analysis, in which it is called failure mode and effect analysis – FMEA) can be used by the designer to select the relevant degradation factors. The selection can be done in two ways:

1. selection based on degradation factors expected at the location of the future structure;

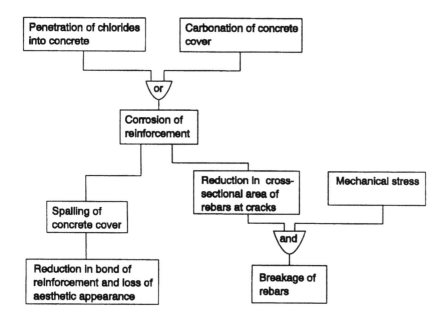

**Figure 7.2** Fault tree for the corrosion of reinforcement.

2. selection based on risk; only degradation factors with a relatively high risk need be considered. In this respect risk is the probability of failure multiplied by the amount of damage caused by degradation.

Mechanisms of great complexity, in which several mechanisms play a part and/or the effect of a mechanism in turn constitutes a degradation factor for another, can be represented with the aid of a tree diagram (Figure 7.2).

Every influence to which the structure is subjected will produce a particular response, depending on the nature of the influence concerned. For every degradation factor there is a response factor in the structure that tends to resist the degradation. For mechanical loading there is bearing capacity, for frost attack there is frost resistance, etc. The internal response factors are related mostly to material properties and geometrical shapes and dimensions. Generally the responses can be classified similarly to the degradation factors as follows:

1. mechanical
2. mechanical–physical
3. physical
4. mechanical–chemical–physical
5. chemical
6. geometrical.

When formulating durability models, human error and other uncertainties should also be considered. This category of influence includes the following groups (Kraker, de Tichler and Vrouwenvelder, 1982):

1. uncertainties of design;
2. errors of communication;
3. uncertainties of manufacture and execution;
4. errors in mathematical and statistical modelling.

If not otherwise treated, uncertainties of design, manufacture and execution may be taken into account as extra scatter in durability models. Omitting some parameters in durability models may also lead to greater scatter. However, gross errors cannot be dealt with in the scatter.

### 7.2.2 Quantification of degradation, performance and service life

The final step in the process of producing durability models is quantification and formulation. Statistical methods and theoretical reasoning are the tools used for these tasks. Simplifications, omitting irrelevant factors and limitation of relevant factors are often necessary actions.

Durability models can be based on empirical or analytical grounds. Empirical models are based on experience and test results. They are developed from results of field surveys and laboratory tests by applying correlative and other statistical methods.

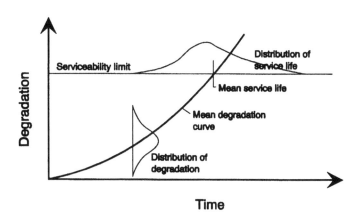

**Figure 7.3** Mean degradation and mean service life.

Analytical models are based on laws of nature and fundamental reasoning. They are created as a thorough analysis of degradation mechanisms and kinetics. Before models can be applied, tests are usually required for determining values for some material properties.

Very often, empirical models represent the viewpoint of engineers, and analytical models that of material scientists. A drawback of empirical models is that mechanisms of influence are poorly understood in models in general. Consequently, any deviation from the limits of the model may not be possible without entailing risk. Analytical methods are based on a deeper understanding of the characteristic features of damage, but their practical importance may be small if the parameters in the model are not measurable or the models cannot otherwise be brought to a level of practical utilization.

Both the empirical and analytical viewpoints should be considered when developing durability models. Models can be considered good when based on an analysis of mechanisms and factors leading to degradation, yet subjected also to laboratory and field tests. A practical requirement for applying models to structural calculation is that they must be reasonably simple and understandable.

When developing an empirical degradation (or performance) model a field investigation is normally necessary. The aim of the research is to find out the effects of the main parameters, especially environmental ones, on degradation over time. Degradation (or performance) is measured in any appropriate quantities or units. Regression analysis and other statistical methods are used for determining the mean degradation curve (Figure 7.3).

The model can then be extended by laboratory tests. To do that it is essential to ascertain that the laboratory test is valid with respect to the degradation factor studied. The effect of different parameters on the rate of degradation is then studied using test series with varying material properties, dimensions, etc. The information gained from these tests, together with the results of field tests, is often sufficient to build up a degradation or performance model with several parameters as follows:

$$\mu(D) = D(x_1, x_2, \dots x_n, t) \tag{7.1}$$

where

$$\mu(D) = \text{the mean degradation,}$$
$$x_1, x_2, x_n = \text{are material, structural, and environmental}$$
$$\text{parameters, and}$$
$$t = \text{the age of the structure.}$$

Performance models can often easily be derived from degradation models, as what they represent in fact is the opposite of degradation. If, for instance, a degradation model shows the depth of carbonation with time, the corresponding performance model would show the non-carbonated part of the concrete cover with time. The non-carbonated portion of concrete cover would in this case be the 'capacity' of the structure, i.e. its performance.

The general form of a performance model resembles that of a degradation model:

$$\mu(P) = (x_1, x_2, \ldots x_n, t) \tag{7.2}$$

where

$$\mu(P) = \text{the mean performance and}$$
$$x_1, x_2, x_n = \text{are material, structural, and environmental parameters, and}$$
$$t = \text{the age of the structure.}$$

At structural level the performance is not usually a linear function of degradation in materials, and more than one degradation factor may be incorporated in a performance model. For example, in a performance model for the bearing capacity of a concrete column, the bearing capacity is related to the cross-sectional areas of concrete and steel. Thus the degradation models of concrete and steel are written into the mathematical functions of cross-sectional areas in which the width of the column and the diameter of the steel are raised to a second power.

The mean service life is usually approximated as the period of time over which the mean degradation reaches the maximum allowable degradation, or over which the mean performance reaches the minimum allowable performance. When the maximum degradation, $D_{\text{max}}$, or the minimum performance, $P_{\text{min}}$, is known the corresponding mean service life can usually be worked out from the degradation or performance models:

$$\mu(t_\text{L}) = t_\text{L}(x_1, x_2, \ldots x_n, D_{\text{max}}) \tag{7.3}$$

or

$$\mu(t_{\mathrm{L}}) = t_{\mathrm{L}}(x_1, x_2, \dots x_n, P_{\min}) \tag{7.4}$$

where $\mu(t_{\mathrm{L}})$ is the mean service life.

If sufficient service life data are available the mean service life can, of course, be modelled directly without first modelling degradation or performance. This is, however, seldom the case. When the mean service life is determined from degradation or performance models in the above manner, the result is not exact but it can often be used as a close approximation of the mean.

### 7.2.3 From deterministic to stochastic durability models

In the stochastic models not only the mean degradation (or performance or service life) is given but also the assumed form of distribution and methods for evaluating the scatter. Instead of a single value a distribution can be obtained for each combination of parameter values by a stochastic model.

In stochastic design deterministic design models are normally used for evaluating the mean degradation. To evaluate the standard deviation a constant coefficient of variation is given. With increasing degradation the standard deviation is also increased.

$$\sigma = \nu\mu \tag{7.5}$$

where

$\sigma =$ the standard deviation of degradation,
$\mu =$ the mean of degradation and
$\nu =$ the coefficient of variation.

Evaluation of the standard deviation can also be based on the differentiation method. In this method the deterministic model of the mean $(X)$ is differentiated with respect to every parameter $(x_i)$ in the model and multiplied by the standard deviation of that parameter. The final standard deviation is then determined as follows:

$$\sigma^2(X) = \sum_{i=1}^{n}\left\{\frac{\partial X}{\partial x_i}\sigma(x_i)\right\}^2 \tag{7.6}$$

where

$\sigma(X) =$ the standard deviation of $X$ (degradation, performance or service life),

$\sigma(x_i) =$ the standard deviation of parameter $x_i$,

$\dfrac{\partial X}{\partial x_i} =$ the partial derivative of $X$ with respect to variable $x_i$, and

$n =$ the number of variables.

The method is exemplified in section 4.3.2.

The Markov chain method can be used to model a typical stochastic degradation process for concrete structures. The method simulates a natural deterioration process starting from the perfect condition and proceeding with gradual and random degradation. Both the scatter and the form of distribution are determined during the process.

The Markov chain method also starts from a known mean curve. However, no parameters or assumptions for the type of distribution are needed. The principle of the Markov chain method is explained in the Appendix (Jiang, Saito and Sinha, 1988; Vesikari, 1995).

Markov chain mathematics has been used in different applications such as network level bridge management systems. A great advantage of the Markov chain method is that it allows linear programming. Thus optimizing analysis can be performed for minimizing repair costs and scheduling repair actions.

# 8

# Durability models for some degradation processes

## 8.1 ABOUT THE MODELS

In the following several examples of durability models for different degradation factors are introduced. They are not presented as the only plausible ones; other models can be used if they meet the formal requirements of durability models and are otherwise appropriate for durability design. To date there is no general consensus concerning durability models.

As the models are intended for use either in deterministic design or in semi-stochastic design with lifetime safety factors, no other information about stochastic properties except the mean is given. The notation $\mu()$ is not presented in the model formulae as it is assumed that the models always express the mean.

Some introductory remarks precede the models where appropriate. These include an explanation of how the incorporation of degradation is attempted in structural design. Problems have been encountered especially in the case of concrete frost attack and other surface weathering, where the degradation of concrete may appear in different forms.

## 8.2 FROST ATTACK

### 8.2.1 Forms of deterioration

By frost attack we mean the gradual weakening or disintegration of concrete surfaces as a result of repeated freezing and thawing. In moist and freezing conditions a decrease in concrete strength occurs and eventual disintegration and complete loss of material is expected near the surface.

Concrete disintegrates as a result of water freezing in its capillary pores. One reason for the damage is the roughly 9% increase in volume that occurs when water freezes. Another reason for the pressure within concrete is the fact that ice crystals tend to grow when kept frozen in a moist environment.

The frost resistance of concrete can be understood as a material property. It is the ability of concrete to withstand repeated cycles of freezing and thawing. The frost resistance depends on other properties of the concrete such as strength, tightness, air content, etc. The rate of disintegration depends, however, not only on the quality of the concrete but also on the aggressiveness of environmental conditions.

The aggressiveness of environmental conditions is exacerbated by deicing salts spread on roads and pavements. Deicing salts not only increase pressures within the concrete, but also diminish its ability to withstand them. A typical feature of frost–salt damage mechanism is gradual scaling of thin concrete layers.

### 8.2.2 Modelling of frost attack

Pure frost damage appears first as a reduction of strength in the edge zone of a concrete structure. Equation (8.1) shows a model for the reduced strength:

$$f_{ck}(d) = f_{ck} \left( 1 - \left( 1 - \left( \frac{d}{H} \right)^n \right) \right) \tag{8.1}$$

where

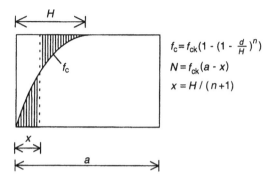

$$f_c = f_{ck}(1 - (1 - \tfrac{d}{H})^n)$$
$$N = f_{ck}(a - x)$$
$$x = H / (n+1)$$

**Figure 8.1** Determination of the apparent loss of concrete.

$f_{ck}(d) =$ the characteristic compressive strength of concrete
at depth $d$,
$f_{ck} =$ the characteristic compressive strength
of undamaged concrete,
$d =$ the depth from the surface,
$H =$ the depth of influence, and
$n =$ an index related to the number of freeze--thaw
cycles (or time).

In structural design, the reduction of strength in concrete at edge zones could be handled as decreased design strength, applied over the whole cross-sectional area of concrete or as reduced dimensions of the cross-section. The latter method is here implemented by introducing the concept of apparent loss of concrete.

Figure 8.1 shows the reduction of strength according to equation (8.1) within the depth of influence, $H$. The value of index $n$ decreases from infinity (at the moment $t = 0$) to 0 (after an infinitely long time). The apparent loss of concrete, $x$, is shown in the figure as calculated. The measure $x$ must be subtracted from the width of the structure, $a$, to obtain the same

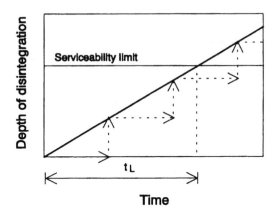

**Figure 8.2**   Model showing concrete loss due to frost.

capacity for the structure with undamaged concrete. Thus even if no real loosening of the concrete takes place, the damage can be taken into account by reducing the dimensions of the structure by the apparent loss of concrete. The apparent loss is the depth of influence divided by the time-related quantity $n + 1$.

Of course, real loosening of concrete also occurs and is highly dependent on other possible forces to which the structure is subjected. Flowing water, ice floats, traffic, etc. are often the final forces causing concrete to loosen. In the absence of such forces, total loss of strength is required before loosening takes place.

The loss of structurally effective concrete as a function of time can be roughly described with a linear model. Figure 8.2 shows such a model in which actual loosening occurs as a series of steps. These can be big or small depending on the presence of chlorides or other possible abrasive forces. The model straightens the steps into a smooth linear process. This is justified by the fact that between steps some apparent loosening takes place in the seemingly whole concrete. The rate of disintegration (meaning the loss of structurally effective concrete in the sense described above) is evaluated from equation (8.2) (Vesikari, 1994).

The effects of curing, ageing, air content and the compressive strength of concrete were studied with frost–salt tests (Matala, 1991). The applicability of the results was also extended to the area of pure frost attack, as the rate of concrete loss can hardly be measured using methods for pure frost resistance (without salt). The environmental factors are based partly on field tests (Vesikari, 1995).

$$r = c_{env}c_{cur}c_{age}a^{-0.7}(f_{ck} + 8)^{-1.4} \qquad (8.2)$$
$$s = rt \qquad (8.3)$$

where

$r =$ the rate of disintegration (loss of structurally effective concrete, mm/year)

$c_{env} =$ the environmental coefficient,

$c_{cur} =$ the curing coefficient,

$c_{age} =$ the ageing coefficient,

$a =$ the air content (%),

$f_{ck} =$ the characteristic cubic compressive strength of concrete at 28 days (MPa), and

$s =$ the loss of structurally effective concrete.

The model formula (8.2) is very conservative regarding the compressive strength of concrete. The reductive effect of compressive strength on the rate of disintegration may be greater than that suggested by the formula. However, at very high compressive strengths extra internal cracking may occur which cannot yet be controlled properly. This is why a careful attitude towards compressive strength has been maintained to date.

The curing coefficient is calculated from

$$c_{cur} = \frac{1}{0.85 + 0.17\log_{10}(d)} \qquad (8.4)$$

where $d$ is the curing time (days).

The ageing coefficient is calculated from

$$c_{\text{age}} = \frac{1}{1 - 0.045p_{\text{sf}} - 0.008p_{\text{sl}} - 0.001p_{\text{fl}}} \tag{8.5}$$

where

$p_{\text{sf}}$ = the proportion of silica fume with respect to the
total weight of binding agent (%),

$p_{\text{sl}}$ = the proportion of blast furnace slag with respect to the
total weight of binding agent, and

$p_{\text{fl}}$ = the proportion of fly ash with respect to the
total weight of binding agent.

The environmental coefficient is evaluated using Table 8.1.

## 8.3 SURFACE DETERIORATION

### 8.3.1 Forms of deterioration

By surface deterioration of concrete structures we mean different types of weathering mechanisms in outdoor conditions, excluding frost attack which is modelled separately. These include temperature and moisture fluctuations, leaching of minerals from concrete, and physical salt weathering.

Daily temperature changes, especially on surfaces exposed to sunshine, cause gradual cracking on concrete edge zones. Wetting and drying cycles with climatic moisture changes also cause slight cracking and small changes in the porosity of concrete. These phenomena are promoted by the possible incompatibility of aggregate quality with the cement–stone matrix.

Water in contact with concrete surfaces causes leaching of concrete minerals. Loss of material is promoted by chemical reactions of these minerals with diluted gases and ions such as

**Table 8.1** Classification of conditions and environmental coefficient values

| Class | Conditions | Value of environmental coefficient |
|---|---|---|
| 1 | Very hard<br>frost, snow, ice, numerous freeze–thaw cycles<br>salt water or deicing salts<br>temperature and moisture variations<br>latitudes $60° \pm 5°$ | 80–160 |
| 2 | Hard<br>frost, snow, ice, numerous freeze–thaw cycles<br>constant contact with water (no chlorides)<br>temperature and moisture variations<br>latitudes $60° \pm 10°$ | 40–80 |
| 3 | Moderate<br>normal outdoor conditions<br>freeze–thaw effect<br>latitudes $60° \pm 10°$ | 20–40 |
| 4 | Favourable<br>no freeze–thaw effect | <20 |

$CO_2$, $SO_2$ and Mg. Flowing water increases the rate of material loss.

Salt weathering is due to a variety of mechanisms associated with the crystallization of salts in the pores of concrete. These mechanisms normally include capillary suction of saline water from the ground or sea, followed by precipitation of salt crystals in pores and cavities while the water is evaporating. An associated mechanism is the expansion and shrinkage of salt crystals as a result of hydration and dehydration leading to cracking and disintegration of concrete.

### 8.3.2 Modelling of surface deterioration

The concept of apparent loss of concrete which was introduced in conjunction with the frost attack model (cf. section 8.2.2) is also applied to the model for surface deterioration. This means that weakening of concrete at the edge zones of a structure is taken into account as a corresponding apparent loss of concrete. The reduction in load-bearing capacity of a structure is evaluated from the loss of effective cross-sectional area of the concrete, not from reduced strength.

Permeability is probably the most influential property related to the durability of structures. The rate of penetration of water and diluted harmful agents depends on permeability, likewise the rate of leaching of important concrete minerals. The permeability of a concrete is related to its compressive strength.

For concretes of moderate and high strength (30–100 MPa) the rate of disintegration (loss of structurally effective concrete) is considered to be constant and inversely proportional to the power of compressive strength. Using the power −3.3 means that doubling the strength reduces the rate of concrete loss to one-tenth. Equation (8.6) is used to evaluate the rate of disintegration (Pihlajavaara, 1994).

$$r = c_{env} c_{cur} f_{ck}^{-3.3} \qquad (8.6)$$

where

$r =$ the rate of disintegration (loss of structurally effective concrete, mm/year)

$c_{env} =$ the environmental coefficient,

$c_{cur} =$ the curing coefficient, and

$f_{ck} =$ the characteristic (cubic) compressive strength of the concrete.

**Table 8.2** Classification of conditions and environmental coefficient values

| Class | Conditions | Value of the environmental coefficient |
|-------|-----------|------------------------------------------|
| 1 | Very hard | 100 000–500 000 |
|   | 'Gulf conditions', latitudes 20° ± 10° | |
|   | marine structures or structures within the capillary rise of saline ground water | |
|   | temperature and moisture variations | |
| 2 | Hard | 10 000–100 000 |
|   | marine structures or structures within the capillary rise of saline ground water | |
|   | latitudes 40° ± 10° | |
|   | temperature and moisture variations | |
| 3 | Normal | 1000–10 000 |
|   | normal outdoor conditions | |
|   | small climatic changes | |
|   | latitudes 40° ± 10° | |
| 4 | Favourable | <1000 |
|   | air continuously dry | |
|   | no sunshine | |

The rate of disintegration is highly dependent on the environmental conditions. The service life of good quality concrete may range from 10 to 10 million years depending on the ambient conditions.

The environmental coefficient is evaluated using Table 8.2.

The formula for the curing coefficient is the same as that used in the model for frost attack (equation (8.4)).

## 8.4 ABRASION OF CONCRETE BY ICE

### 8.4.1 Forms of deterioration

A concrete offshore structure in arctic conditions is subject to various degradation factors. Based on their effect these can be classified as mechanical, physical or chemical. Chemical changes due to sea water include penetration of chlorides and dissolution of lime. Physical changes due to freezing and thawing include microcracking. However, the final cause for detachment of concrete in arctic conditions is almost always abrasion by ice.

### 8.4.2 Modelling of abrasion by ice

The abrasion mechanism due to crushing ice sheets against the concrete surface is of three kinds; abrasion of cement stone (Figure 8.3(a)), abrasion of cement stone + loosening of

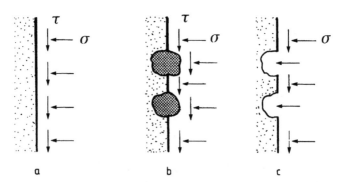

**Figure 8.3**  Abrasion mechanisms. (Redrawn from S. Huovinen, *Publications No. 62*, published by Technical Research Centre of Finland, 1990.)

protruding aggregate stones (Figure 8.3(b)), and abrasion of cement stone when the bond strength between larger aggregate stones and the cement stone is so weak that the stones loosen during the first ice impact (Figure 8.3(c)) (Huovinen, 1990).

The abrasion depth of concrete in arctic sea structures can be calculated as the sum total of the abrasion depth of cement stone measured in icebreaker tests at sea, and the loosening of aggregate stones from the concrete surface. The abrasion rate for cement stone is obtained from the results of icebreaker tests as follows:

$$b = \frac{3}{f_{ck}} s \ \text{(mm/km) (mechanism a)} \qquad (8.7)$$

where

$$s = \text{the movement of the ice sheet (km) and}$$
$$f_{ck} = \text{the characteristic compressive cubic strength of}$$
$$\text{the concrete (MPa).}$$

The total abrasion depth can be calculated with the formula

$$\text{ABR} = \sum_{i=1}^{n} a_i \frac{\log n_s}{\log n_1} R_i + (1 - \sum a_i)b \quad \text{(mechanism b)} \qquad (8.8)$$

where

$$a_i = \text{is the proportional volume of aggregate stones of}$$
$$\text{radius } R_i \text{ in the concrete,}$$
$$n_s = \text{the number of ice impacts during ice sheet movements,}$$
$$n_1 = \text{the number of ice impacts when the aggregate stone is}$$
$$\text{loosening } (L_{cr}/R = 1) \text{ and}$$
$$b = \text{the abrasion rate of cement stone (mm).}$$

Abrasion as a function of ice sheet movement calculated using equation (8.8) is presented graphically in Figure 8.4. The abrasion

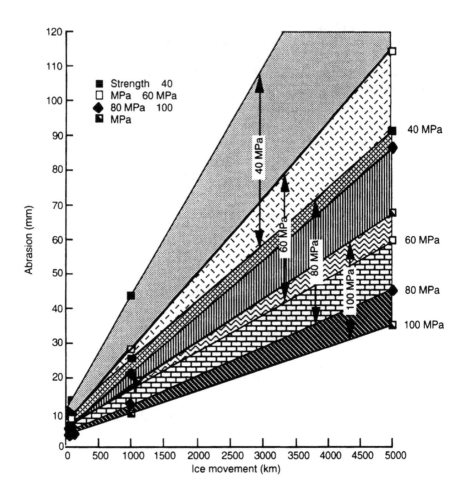

**Figure 8.4**   Abrasion of concrete for compressive strengths $f_{ck}$ = 40, 60, 80 and 100 MPa as a function of ice sheet movement. (Redrawn from S. Huovinen, *Publications No. 62*, published by Technical Research Centre of Finland, 1990.)

diagrams are valid when the aggregate distribution of concrete is normal. In addition to the compressive requirement $f_c$ = 40, 60, 80, 100 MPa it is presupposed that the tensile strength of concrete $f_{ct}$ is at least 10% of the compressive strength and the bond strength between aggregate stones is at least 90% of the tensile strength of concrete (Huovinen, 1990). Figure 8.4 shows that the latter term in equation (8.8) dominates with increasing ice movement.

If the bond between aggregate stones and the cement stone has deteriorated under freeze–thaw cycling the abrasion depth can be calculated with the formula

$$\text{ABR} = \frac{1}{(1 - \sum a_i) f_{ck}} \cdot \frac{3}{s} \quad \text{(mechanism c)} \tag{8.9}$$

where

$s =$ is the movement of the ice sheet (km) and

$\sum a_i =$ the total proportional volume of aggregate stones in concrete.

In the structural design of concrete structures the following approximations for the rate of abrasion can be used:

1. when aggregate stones are not loosening due to frost attack:

$$\frac{d(\text{ABR})}{dt} = \frac{3p'}{f_{ck}} v \quad \text{mm/year} \tag{8.10}$$

2. when aggregate stones are loosening due to frost attack:

$$\frac{d(\text{ABR})}{dt} = \frac{3}{p' f_{ck}} v \quad \text{mm/year} \tag{8.11}$$

where

$v =$ the movement of the ice sheet (km) in 1 year,

$p' =$ the total proportional volume of cement stone in concrete including aggregates up to $\phi = 4\,\text{mm}$ ($\sim$ 0.4--0.6 depending on the proportioning of concrete), and

$f_{ck} =$ the characteristic compressive cubic strength of concrete (MPa).

Ice sheet movement may be as much as 10 000 km a year depending on the velocity of wind and sea currents.

## 8.5 CORROSION OF REINFORCEMENT

### 8.5.1 Concrete as a protective material for reinforcement

The co-operation of concrete and steel is based partly on the fact that concrete gives the reinforcement both chemical and physical protection against corrosion. The chemical effect of concrete is attributed to its alkalinity, which causes an oxide layer to form on the steel surface. This phenomenon is called passivation, as the oxide layer prevents propagation of corrosion. The concrete also provides the steel with a physical barrier against agents that promote corrosion, such as water, oxygen and chlorides.

In normal outdoor concrete structures, corrosion of reinforcement takes place only if changes occur in the concrete surrounding the steel. The changes may be physical, such as cracking and disintegration, exposing part of the steel surface to open air and leaving it without the physical and chemical protection of concrete.

Chemical changes also take place in the concrete surrounding the reinforcement; the most important are the following:

1. carbonation of concrete due to carbon dioxide in air;
2. penetration of aggressive anions, especially chlorides, into concrete.

Carbonation is the reaction of carbon dioxide (in air) with hydrated cement minerals in concrete. This phenomenon occurs in all concrete surfaces exposed to air, resulting in lowered pH in the carbonated zone. In carbonated concrete the protective passive film on steel surfaces is destroyed and corrosion is free to proceed. The effect of chlorides is not based on the decrease in pH but on their ability otherwise to break the passive film.

## 8.5.2 Modelling of the corrosion of reinforcement

Two limit states can be identified with regard to service life (Figure 8.5):

1. The service life ends when the steel is depassivated. This rule is usually applied to all chloride-induced corrosion as the local attack penetration rate is still not safely quantified and uncertainties concerning the propagation period are therefore high. Thus the service life is limited to the initiation period only (time for the aggressive agent to reach the steel and induce depassivation).

   This rule is also applied to all prestressing steels. The tensile stress of tendons is normally so high that no reduction in the cross-sectional area is permissible and as a result of surface corrosion there is a risk of stress corrosion cracking.

   In the cases where no corrosion is allowed the following formula for service life can be used:

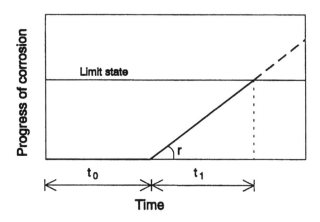

**Figure 8.5** Determination of service life with respect to corrosion of reinforcement.

$$t_L = t_0 \qquad\qquad (8.12)$$

where

$$t_L = \text{the service life, and}$$
$$t_0 = \text{the initiation time of corrosion.}$$

2. The limit state is based on cracking of the concrete cover due to oxides generated during corrosion. In this case the service life includes a certain propagation period of corrosion during which the cross-sectional area of steel is progressively decreased, the bond between steel and concrete is reduced and the effective cross-sectional area of concrete is diminished due to spalling of the cover. This approach is applied in cases where generalized corrosion is developing due to carbonation.

   The service life based on cracking of the concrete cover is defined as the sum of the initiation time of corrosion and the time for cracking of the concrete cover to a given limit.

$$t_L = t_0 + t_1 \qquad\qquad (8.13)$$

where $t_1$ is the propagation time. The propagation time $t_1$ ends when a certain maximum allowable loss of the cross-sectional area or loss of bond or crack width is reached. These values will depend upon the particular detailing and geometry of each element.

At cracks, originating from the beginning of service life, the initiation time $t_0$ is much shorter than in an uncracked cover or even $t_0 = 0$. In this case it may be written:

$$t_L = t_1 \qquad\qquad (8.14)$$

where $t_1$ is the free corrosion time.

Models for estimating $t_0$ and $t_1$ are presented below. When developing these models the assumption has been made that concrete surfaces are free from coatings and sealants.

## 8.5.3 Initiation time of corrosion

### (a) Chloride-induced corrosion

The commonest sources of chlorides are sea water (marine environments) and deicing salts. Admixed chloride is not considered here.

As a result of chloride penetration a gradient develops near the concrete surface. The time at which the critical chloride content (threshold value) reaches the steel surface and depassivates it, can be regarded as the initiation time of corrosion. The gradient of chloride content is often described by an error function model which fulfils the condition of Fick's second law of diffusion:

$$C_x = C_s \left( 1 - \mathrm{erf} \left( \frac{x}{2(Dt)^{1/2}} \right) \right) \tag{8.15}$$

where

$C_x =$ the chloride content at depth $x$,
$C_s =$ the chloride concentration at the concrete surface,
$x =$ the depth from the surface of the structure,
$D =$ the diffusion coefficient, and
$t =$ time.

The initiation time for corrosion is obtained from the formula

$$C_{th} = C_s \left( 1 - \mathrm{erf} \left( \frac{c}{2(Dt_0)^{1/2}} \right) \right) \tag{8.16}$$

where

$C_{th} =$ the critical chloride content,
$c =$ the concrete cover, and
$t_0 =$ the initiation time of corrosion.

This formula may be simplified by using a parabola function:

$$C_x = C_s \left( 1 - \frac{x}{2(3Dt)^{1/2}} \right)^2 \tag{8.17}$$

The formula for the initiation time of corrosion can then be written in the following form:

$$t_0 = \frac{1}{12D} \left( \frac{c}{1 - (C_{th}/C_s)^{1/2}} \right)^2 \tag{8.18}$$

Many standards require threshold values not higher than 0.4% (Cl⁻) by weight of cement for reinforced concrete and 0.2% for prestressed concrete. This corresponds approximately to 0.05–0.07 by weight of concrete (0.025–0.035 for prestressed concrete).

Concerning values of $C_s$, field experience has shown this quantity to be time dependent at early ages but to tend towards a maximum after a number of years. For the sake of calculation it is usually considered constant. Normal values may be about 0.3–0.4 by weight of concrete.

The coefficient of diffusion is roughly $10^{-7}$–$10^{-8}$ cm²/s.

### (b) Stress corrosion cracking

Fortunately the phenomenon of stress corrosion cracking is uncommon. It may develop in prestressing wires subjected to corrosive agents, leading to brittle fracture with almost no loss in cross-sectional area. Stress corrosion is incubated in very small surface cracks.

Local steel depassivation is needed to produce surface cracks in which the stress corrosion can incubate. Therefore protecting the prestressing steels from aggressive agents is crucial to their service life, which is always limited to the initiation time of corrosion. As regards intrusion of chlorides the calculation rules presented in section 8.5.3(a) can also be applied to prestressing steels.

### (c) Carbonation-induced corrosion

Carbon dioxide in the air penetrates concrete, neutralizing its alkaline substances and producing a carbonation front which advances towards the interior. When this carbonation front reaches the reinforcement, the passive film on the steel becomes unstable and dissolves, enabling generalized corrosion to occur. The initiation time of corrosion is defined as the period of time needed for complete carbonation of the concrete cover.

The rate of carbonation is usually assumed to be related to the square root of time:

$$d = K_c t^{1/2} \tag{8.19}$$

where

$$
\begin{aligned}
d &= \text{the depth of carbonation at time } t, \\
K_c &= \text{the carbonation coefficient and} \\
t &= \text{time or age.}
\end{aligned}
$$

The initiation time of corrosion can be determined as follows:

$$t_0 = \left(\frac{d}{K_c}\right)^2 \tag{8.20}$$

The carbonation coefficient depends on the strength of concrete, binding agents, cement content and environmental conditions (humidity and temperature). There are several formulae for modelling the carbonation rate. Some of them are analytical, others empirical.

Based on Fick's first law the following expression can be derived for the depth of carbonation (Schiessl, 1976):

$$x = \left(\frac{2D_c(C_1 - C_2)}{a}t\right)^{\frac{1}{2}} \tag{8.21}$$

where

$x =$ is the carbonation depth (m),

$a =$ the amount of alkaline substance in the concrete,

$D_c =$ the effective diffusion coefficient for $CO_2$ at a given moisture distribution in the pores ( $m^3/s$),

$C_1 - C_2 =$ the concentration difference of $CO_2$ between air and the carbonation front ($kg/m^3$), and

$t =$ time.

This calculation procedure has been extended by Bakker (1993) to cases of fluctuating wetting and drying cycles. During wet conditions the carbonation front cannot progress. During dry conditions moisture evaporates and makes possible progression of the carbonation front.

According to Bakker the time $t$ in equation (8.21) is substituted by $t_{eff}$ which is determined as follows:

$$t_{eff} = \left( t_{d1} + t_{d2} - \left( \frac{x_1}{B} \right)^2 + t_{d3} + \ldots t_{dn} - \left( \frac{x_{n-1}}{B} \right)^2 \right) \qquad (8.22)$$

$$B = \left( \frac{2D_v(C_3 - C_4)}{b} \right)^{1/2} \qquad (8.23)$$

where

$x_n =$ the carbonation depth after the $n$th wetting and drying cycle (m),

$t_{dn} =$ the length of the $n$ th drying period,

$D_v =$ the effective diffusion coefficient for water vapour at a given moisture distribution in the pores ($m^2/s$),

$C_3 - C_4 =$ the moisture difference between air and the evaporation front ($kg/m^3$), and

$b =$ the amount of water to evaporate from the concrete ($kg/m^3$).

If the drying and wetting periods are of equal length the time elapsed after $n$ cycles is

$$t_n = nt_d + (n-1)t_w \qquad (8.24)$$

where

$t_w =$ the length of the wetting periods and

$t_d =$ the length of the drying periods.

A theoretical model based on the theory of 'moving boundaries' has been presented by Tuutti (1982). The theory deals with diffusion processes in non-steady-state conditions where $CO_2$ reacts with concrete in such a way that concrete serves as a sink for $CO_2$. Another theoretical model for the combined effects of frost attack and carbonation has been presented by Fagerlund, Somerville and Tuutti (1994).

Experimental models for evaluating the depth of carbonation have been presented by Häkkinen and Parrot. According to Häkkinen (1993) the depth of carbonation is determined by equation (8.20), the coefficient of carbonation being determined as follows:

$$K_c = c_{env} c_{air} a f_{cm}^b \qquad (8.25)$$

where

$c_{env} =$ the environmental coefficient,

$c_{air} =$ the air content coefficient,

$f_{cm} =$ the mean (cubic) compressive strength of concrete (MPa), and

$a, b =$ parameters depending on the binding agent.

Instead of the mean compressive strength, the characteristic strength can be used by applying the following relationship (CEB, 1988):

$$f_{cm} = f_{ck} + 8 \qquad (8.26)$$

Tables 8.3 and 8.4 show values for the environmental and air content coefficient respectively. The parameters $a$ and $b$ in equation (8.25) are presented in Table 8.5 (Häkkinen, 1991).

**Table 8.3** Environmental coefficient for determination of the carbonation rate

| Environment | $c_{env}$ |
|---|---|
| Structures sheltered from rain | 1 |
| Structures exposed to rain | 0.5 |

**Table 8.4** Air content coefficient for determination of the carbonation rate

| Air porosity | $c_{air}$ |
|---|---|
| Not air entrained | 1 |
| Air entrained | 0.7 |

**Table 8.5** Parameters $a$ and $b$ (Source: T. Häkkinen, *Research. Notes 750*, published by Technical Research Centre of Finland, 1991)

| Binder | $a$ | $b$ |
|---|---|---|
| Portland cement | 1800 | −1.7 |
| Portland cement + fly ash 28% | 360 | −1.2 |
| Portland cement + silica fume 9% | 400 | −1.2 |
| Portland cement + blast furnace slag 70% | 360 | −1.2 |

According to Parrot (1992) the depth of carbonation is determined on the basis of the oxygen permeability of concrete:

$$d = \frac{64K^{0.4}t^n}{c^{0.5}} \tag{8.27}$$

where

> $K =$ the oxygen permeability of concrete at 60% RH,
>
> $t =$ time,
>
> $c =$ the alkaline content in the cement, and
>
> $n =$ the attenuation factor (root power).

## 8.5.4 Propagation period

### (a) General rule

Corrosion begins when the passive film is destroyed as a result of falling pH due to carbonation, or as a result of the chloride content rising above the threshold close to the reinforcement. The volume of corrosion products is many times that of the original metal. The greater need for volume causes tensile stress in concrete around the steel bar, leading to cracking or spalling of the concrete cover.

When corrosion develops three main phenomena appear:

1. a decrease in the steel cross section;
2. a decrease in the steel/concrete bond;
3. cracking of the concrete cover and therefore a decrease in the concrete load-bearing cross-section.

To determine the length of service life the critical threshold value of the load-bearing capacity has to be defined as related to the aforementioned distressing phenomena. This critical threshold can often be expressed as the critical loss of bar radius provoked by corrosion and, therefore, the propagation period may be quantified in the following manner (Alonso and Andrade, 1993):

$$t_1 = \frac{\Delta R_{max}}{r} \tag{8.28}$$

where

$\quad t_1 =$ the propagation time of corrosion (years),

$\Delta R_{max} =$ the maximum loss of radius of the steel bar, and

$\quad r =$ the rate of corrosion.

## (b) Cracking time of the concrete cover

In the case of generalized corrosion the critical loss of bar radius is based on the cracking of the concrete cover. The propagation (cracking) time can be approximated by the following formula (Siemes, Vrouwenvelder and van den Beukel, 1985):

$$t_1 = 80 \frac{C}{Dr} \tag{8.29}$$

where

$\quad C =$ the thickness of the concrete cover (mm),

$\quad D =$ the diameter of the rebar (mm), and

$\quad r =$ the rate of corrosion in concrete ($\mu$m/year).

The rate of corrosion in concrete depends strongly on the ambient conditions. Important environmental factors are relative humidity and temperature. The rate of corrosion of reinforcement in concrete can be evaluated using the following formula:

$$r = c_T r_o \tag{8.30}$$

where

$\quad c_T =$ the temperature coefficient,

$\quad r_o =$ the rate of corrosion at $+20°$ C.

**Table 8.6** Rate of corrosion in carbonated and chloride-contaminated concrete (anodic areas)

| Relative humidity (%) | Carbonated concrete ($\mu$m/year) | Chloride-contaminated concrete ($\mu$m/year) |
|---|---|---|
| 99 | 2 | 34 |
| 95 | 50 | 122 |
| 90 | 12 | 98 |
| 85 | 3 | 78 |
| 80 | 1 | 61 |
| 75 | 0.1 | 47 |
| 70 | 0 | 36 |
| 65 | 0 | 27 |
| 60 | 0 | 19 |
| 55 | 0 | 14 |
| 50 | 0 | 9 |

Primary factors that affect the rate of corrosion in concrete at +20 °C are the relative humidity of air (or concrete) and the chloride content. Other factors such as the water–cement ratio and the type of cement may also have some influence. The values of corrosion rate in anodic areas of reinforcement presented in Table 8.6 can be taken as approximate averages. They are determined on the basis of experimental data reported by Tuutti (1982).

The moisture content of concrete surrounding the reinforcing steels is a complex mixture of various climatic and structural effects. The equilibrium relative humidity of concrete in aerial conditions is affected by annual and daily variations of the relative humidity of air, condensation of moisture on the surfaces, rain, splash and melting water, density of concrete and depth from the surface (concrete cover).

The chloride content also has a great influence on the moisture

**Table 8.7** Temperature coefficients and evaluated rates of corrosion for some cities in Europe

| City | $c_T$ | Rate of corrosion ($\mu$m/year) | |
| --- | --- | --- | --- |
| | | Exposed to rain | Sheltered from rain |
| Sodankylä (northern Finland) | 0.21 | 11 | 2.5 |
| Helsinki | 0.32 | 16 | 4 |
| Amsterdam | 0.47 | 24 | 6 |
| Madrid | 0.73 | 37 | 9 |

content and the rate of corrosion in concrete. However, the propagation time is normally fully omitted if chlorides are present. The data for chloride-contaminated concrete in Table 8.6 are given mainly for comparison.

The average relative humidity in structures exposed to rain can be evaluated at about 95% (unless the frequency of rains is extremely low) and for structures completely sheltered from rain, about 90%. Consequently the rate of corrosion in carbonated concrete at 20°C would be about 50 $\mu$m/year in structures exposed to rain and about 12 $\mu$m/year in structures sheltered from rain.

The temperature coefficients determined on the basis of the findings by Tuutti (1982) and average daily temperatures for some European cities are presented in Table 8.7. Rates of corrosion evaluated from equation (8.30) are also listed. The effect of direct sunshine on the surface temperatures of structures has not been considered in Table 8.7. This effect may, however, be considerable. Local microclimatic features should be taken into account when evaluating the rate of corrosion.

It is well known that the rate of corrosion slows gradually with time. However, as few data are available concerning this phenomenon, a constant corrosion rate is recommended in durability design.

## (c) Propagation time of corrosion at cracks

If the concrete cover is cracked from the beginning (due to shrinkage, mechanical stress, etc.) and the crack width is larger than 0.1–0.3 mm, corrosion normally starts without any initiation period. If the steel bars are exposed all around, even corrosion is expected on all sides.

A constructor may set a limit for the minimum diameter of steel bars or the maximum depth of corrosion correspondingly. This may depend on the type of reinforcement – main reinforcement, transverse reinforcement, stirrups, etc. – and the actual stresses in steel bars. No corrosion in prestressing tendons is permissible.

The propagation time at cracks is calculated from the following formulae:

$$t_1 = \frac{s_{max}}{r} \tag{8.31}$$

$$t_1 = \frac{D - D_{min}}{2r} \tag{8.32}$$

where

$t_1 = $ the propagation time of corrosion at a crack,

$r = $ the rate of corrosion at a crack,

$s_{max} = $ the maximum allowable depth of corrosion, and

$D_{min} = $ the minimum diameter of the steel bar.

The rate of corrosion in cracks represents an extremely complicated problem which is not yet fully understood. In the absence of more precise data the assumption that the average corrosion rate is of the same order of magnitude as in uncracked concrete is applied. Accordingly the following values for the mean corrosion rates are recommended in the calculations (Andrade *et al.*, 1994):

1. when the only aggressive action is carbonation:

$$\text{RH} = 90\text{--}98\% \Rightarrow \text{corrosion rate} = 5\text{--}10 \ \mu\text{m/year}$$
$$\text{RH} < 85\% \Rightarrow \text{corrosion rate} \leq 2 \ \mu\text{m/year}$$

2. in chloride-contaminated environments:

$$\text{RH} = 100\% \Rightarrow \text{corrosion rate} \leq 10 \ \mu\text{m/year}$$
$$\text{RH} = 80\text{--}95\% \Rightarrow \text{corrosion rate} = 50\text{--}100 \ \mu\text{m/year}$$
$$\text{RH} < 70\% \Rightarrow \text{corrosion rate} \leq 2 \ \mu\text{m/year}$$

# 9

# Summary and conclusions

## 9.1 DURABILITY DESIGN IN THIS DESIGN GUIDE

This design guide is in principle based on well-known and accepted stochastic design methods long used in the mechanical design of concrete structures. Here these methods are extended to cover the dimension of time, allowing time-related degradation processes of materials and structures to be taken into account. A new requirement, the target service life, has been introduced in structural design.

This guide is not a formal design code. Rather, it shows in terms of principles, rules and examples how the durability design of concrete structures can be performed and what can be gained by such a design. The purpose of the work has been to bring together the achievements of material research of concrete structures, and to transfer the main results as 'durability models' to the art of structural design.

Structural durability design does not only mean the specification of material properties or the dimensioning of some parts of concrete structures, such as material specifications and thickness of the concrete cover, but it also covers the mechanical design of structures. This is possible by incorporating durability models for concrete and steel into the basic design formulae of load-bearing capacity. In this way, performance models are produced for structural level, showing the reduction of load-bearing capacity with time.

To take the stochasticity of service life into account a new parameter, the lifetime safety factor, has been introduced. The meaning of the lifetime safety factor is the same as that of load safety factors and material safety factors in traditional design. It provides the necessary safety margin against falling short of the target service life. Special statistical methods have been used in the determination of lifetime safety factors.

The guide also gives examples of possible durability models for structural design. Models related to frost attack, surface deterioration and abrasion by ice are presented for concrete; models for carbonation and chloride-induced corrosion are presented for reinforcing steel.

## 9.2 NEEDS FOR FURTHER DEVELOPMENT OF DURABILITY DESIGN

Only a small part of all durability research has so far focused on methodological problems. Further studies on design methods for concrete structures subjected to time-related degradation processes are still very much needed.

More accurate estimates for the scatter of service life in different working environments are needed in order to optimize the values of lifetime safety factors.

The required safety of structures during service life, and the related failure probability of not reaching the required service life should be further studied. Lifetime safety factors for concrete structures should be agreed upon internationally in norms and standards.

The methods of durability design should also be applied to special types of concrete structures. Analysis of the design methods under diverse action effects still needs to be studied in depth. Not only are 'first level' cross-sectional studies needed, but also 'second level' investigations of the performance of the whole structure, focusing on deformations, sliding and buckling effects, etc.

The durability models chosen for this guide are presented as examples. No doubt there is a need for further development of many models. The accuracy of the models can be improved by appropriate field and laboratory research projects. For some degradation factors complete modelling work has yet to be done.

Additionally the criteria of the service life limit state of each degradation process should be further analysed and agreed upon for international praxis.

The interaction of degradation factors has not been studied in this report. The qualitative and quantitative analysis of different forms of interaction, and the methods of incorporating such interactive processes into durability design, is also a challenge for future research.

This design guide is but one link in the chain of development of the durability design of concrete structures. Its purpose has been to outline the principles of durability design and to reveal, by means of examples, the possibilities and benefits of such a design. However, a more formal durability guide is required for structural engineers, in order to firmly establish durability design. This would be especially important for structures in aggressive environments and for structures at great economic and environmental risk.

The durability design methodology also opens possibilities for applications to the rehabilitation and repair of structures. In such application the performance capacity is increased at the time of repair, from which point of time a new period of service life begins. This requires application of the design procedure, and development of degradation models through theoretical and experimental research of different types of repair methods.

# Appendix

# Stochastic modelling of degradation by the Markov chain method

The degradation of structures is expressed in terms of discrete condition states. In principle the number of states is not restricted, but the calculations are easier if the number of states is not too large. As an example a degradation index scale of 0, 1, 2, 3, 4 is used, each index corresponding to a state. The degradation index 0 represents the best and 4 the poorest condition. In general we give the interpretation for the five degradation indices listed in Table A.1. Degradation index 3 defines the limit state. Service life is the age at which this state is reached.

The amounts of structures at each state and at a certain age, $t$, is expressed as a damage index distribution, $Q(t)$. The amounts can be given in any quantity: $m^2$, m, pieces or percentages.

Changes in condition are expressed as stochastic transition probabilities from one state to another. The transition probabilities are given in a transition probability matrix (or simply transition matrix). Transitions normally mean changes occurring within the space of 1 year (but may be other). Changes after $n$ years can be predicted by multiplying the initial degradation index distribution, $Q(0)$, by the transition matrix $n$ times, as shown in Figure A.1. If we assume that all structures start off in perfect condition, the initial degradation index distribution is of the form | 1, 0, 0, 0, 0 |, meaning that all structures are at the degradation index 0. It is important to note that the transition probability matrix does not change throughout the process. The transition probabilities are not dependent on time.

**Table A.1**  Interpretation of degradation indices

| Degradation index | State of definition |
| --- | --- |
| 0 | Initial state, no degradation |
| 1 | $\frac{1}{3}$ limit state |
| 2 | $\frac{2}{3}$ limit state |
| 3 | Limit state |
| 4 | Post-limit state |

For the transition probabilities the following assumptions are made.

1. The condition of structures cannot be improved during the process.
2. The condition state can either remain the same or shift to the next state within 1 year (transition period).

With the first assumption the structures are never repaired, nor is so-called 'self healing' possible. Because of this assumption, all transition probabilities below the diagonal probabilities are zero. The second assumption can be considered reasonable and moderate for normal deterioration processes of concrete structures in outdoor conditions. Due to this assumption, all probabilities above those next to diagonal ones are zero.

As the structures must either remain at the same state or drop to the next one within a year, the sum of the probabilities of remaining (diagonal elements) and dropping to the next state (elements next to the diagonal ones) must be 1. Thus the probabilities of dropping to the next state can be calculated by subtracting the diagonal probability values from 1. Consequently only the diagonal probabilities of the matrices are unknown parameters.

Since degradation index 4 is the highest possible, structures at that level must always keep their degradation index, and the corresponding transition probability is always 1 (at the lower right corner of the matrix).

The assumed form of the transition matrix is thus as follows:

$$P = \begin{vmatrix} p_1 & 1-p_1 & 0 & 0 & 0 \\ 0 & p_2 & 1-p_2 & 0 & 0 \\ 0 & 0 & p_3 & 1-p_3 & 0 \\ 0 & 0 & 0 & p_4 & 1-p_4 \\ 0 & 0 & 0 & 0 & 1 \end{vmatrix}$$

Let us consider that the probabilities $p_1$, $p_2$, $p_3$ and $p_4$ have numerical values (between 0 and 1). The degradation index distribution for each year, $Q(t)$, is obtained by multiplying the degradation index distribution of the previous year by the transition matrix $P$ (starting from the initial degradation index distribution, see Figure A.1). The result is shown in Figure A.2.

The mean of the degradation index distribution at each year, $E(t, P)$, is obtained by multiplying the scale vector, $R = |0,1,2,3,4|$, by the degradation index distribution (vector multiplication).

$$Q(t) = Q(t-1) * P \tag{A.1}$$
$$E(t,P) = Q(t) * R \tag{A.2}$$

The principle now is to compare the mean curve obtained with the reference degradation curve, which is assumed to be known. The values for $p_1$, $p_2$, $p_3$ and $p_4$ are selected to give the best fit to the reference degradation curve. Mathematically the principle is formulated by minimizing the sum of yearly deviations between the reference degradation curve and the Markov estimation for the degradation curve:

$$\min \text{SUMD} = \sum_{t=1}^{N} |S(t) - E(t,P)| \tag{A.3}$$

where

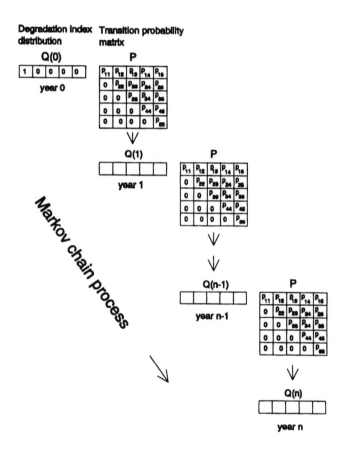

**Figure A.1** Principle of the Markov chain method.

SUMD = the sum of deviations at each year,

$\quad$ N = the number of years within service life (years between states 3 and 4 are not interesting and are omitted),

$\quad$ P = the transition matrix with unknown probability elements, $P_i$,

$\quad$ S(t) = the value of the degradation index of the reference degradation curve at year $t$, and

$\quad$ E(t, P) = the mean of the degradation index distribution calculated by the Markov chain method at year $t$.

In other words, the combination of values $p_i$ that yields the minimum for equation (A.3) is sought. The task is solved as a programming problem.

Let us assume that the reference degradation curves (showing the mean degradation as a function of age) are of three optional types:

1. linear
2. quadratic and
3. square root type.

The curves start from the origin and cross the serviceability limit at a certain critical age, which we call the mean service life. By these two fixed points the curves are fully defined. The types of degradation curve are presented in Figure A.3.

When the reference degradation curve is given, the matrix model is also defined by the process described earlier. As a result we obtain the values for the probabilities $p_1$, $p_2$, $p_3$ and $p_4$. In addition, all necessary information about propagation of the stochastic degradation process is obtained. The probability functions and probability density functions of service life corresponding to the above-mentioned reference curves and the mean service life of 60 years are presented in Figure A.4.

The tractability of the Markov chain mean curve is best with respect to the square root reference curve and worst with respect to the quadratic reference curve. This is due to the fact that the

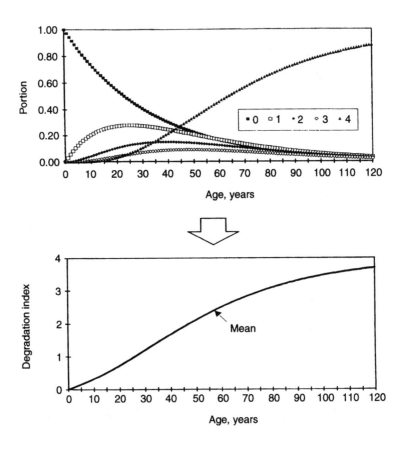

**Figure A.2**  Degradation index distributions for each year,
calculated by the Markov chain method, and curve for the mean
degradation index.

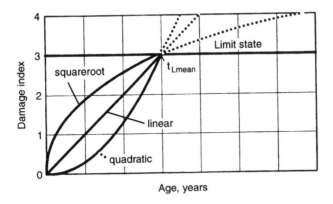

**Figure A.3** Types of reference degradation curve.

Markov mean curve is bowed horizontally as the age approaches infinity. The mean can never be greater than the greatest damage index of 4.

For the above reason, comparison between the Markov chain mean curve and reference curves is done only within damage indices 1–3. As damage index 3 is determinative for the service life the form of the reference curve beyond it is not important. The portions of damage indices 3 and 4 are totalled in the probability functions of service life.

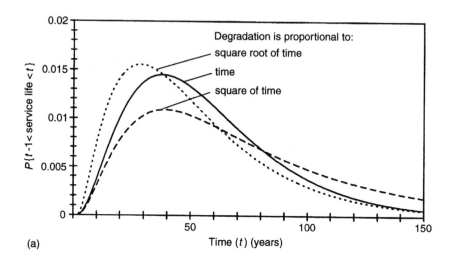

(a)

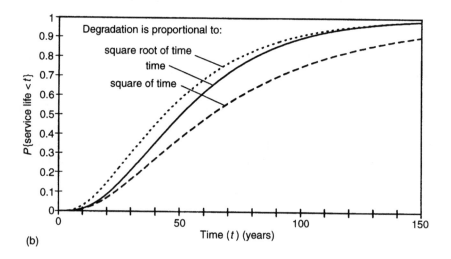

(b)

**Figure A.4**   (a) Probability density functions and (b) (cumulative) probability functions of service life determined by the Markov chain method.

# Bibliography

Alonso, C. and Andrade, C. (1993) Life time of rebars in carbonated concrete. Proceedings of the 10th European Corrosion Congress, Barcelona, *Progress in the Understanding and Prevention of Corrosion*, Vol. 1, pp. 634–41.

Andrade, C., Alonso, C., Gonzales, I.A. and Rodriguez, J. (1989) Remaining service life of corroding structures. *Proceedings of the IABSE Symposium Durability of Structures*, Lisbon, pp. 359–64.

Andrade, C., Alonso, M. C., Pettersson, K., Somerville, G. and Tuutti, K. (1994) The practical assessment of damage due to corrosion. *Proceedings of Int. Conf. Concrete across Borders 1994*, Danish Concrete Association, Odense, pp. 337–50.

Bakker, R. (1993) Model to calculate the rate of carbonation in concrete under different climatic conditions. May. Paper no. 104 – CEN TC 104/WG1/TG1/Panel 1. Unpublished.

BSI (1992) *BS 7543 Guide to Durability of Buildings and Building Elements, Products and Components*. British Standards Institution, London, 48 pp.

CEB (1988) *CEB–FIP Model Code 1990*. First predraft 1988. Comité Euro-International du Béton, Lausanne. U.S. (CEB Bulletin d'information 190a).

CEN (European Committee for Standardization) (1994) *ENV 1991 – 1. Eurocode 1: Basis of Design and Actions on Structures. Part 1: Basis of Design*, Brussels.

Dhir, R. K., Jones, M. R. and Ahmed, H. E. H. (1991) Concrete durability: estimation of chloride concentration during design life. *Magazine of Concrete Research*, 43(154), 37–44.

Fagerlund, G., Somerville, G. and Tuutti, K. (1994) The residual service life of concrete exposed to the combined effect of frost attack and reinforcement corrosion. *Proceedings of Int. Conf. Concrete across Borders 1994*, Danish Concrete Association, pp. 351–64.

Huovinen, S. (1990) Abrasion of concrete by ice in arctic sea structures (Doctoral thesis). Technical Research Centre of Finland, Espoo. Publications No. 62.

Häkkinen, T. (1991) Influence of cementing materials on the permeability of concrete. Technical Research Centre of Finland, Espoo. *Research Notes 750*, 84 pp. + app. 27 pp.

Häkkinen, T. (1993) Influence of high slag content on the basic mechanical properties and carbonation of concrete. Technical Research Centre of Finland, Espoo. *Publications 141*, 98 pp. + app. 46 pp.

Jiang, Y., Saito M. and Sinha K. C. (1988) Bridge performance prediction model using Markov chain. *Transportation Research Record 1180*, pp. 25–32. Transportation Research Board Business Office, Washington.

Kasami, H., Izumi, I., Tomosawa, F. and Fukushi, I. (1986) Carbonation of concrete and corrosion of reinforcement in reinforced concrete. *First Joint Workshop on Durability of Reinforced Concrete, Australia–Japan Science and Technology Agreement*, Tsukuba, Japan, 30 September–2 October, 12 pp.

Kraker, A., de Tichler, J. W. and Vrouwenvelder, A.C.W.M. (1982) Safety, reliability and service life of structures. *Heron* **27(1)**, 85 pp.

Masters, L. W. and Brandt, E. (1989) Systematic methodology for service life prediction of building materials and components. Report of the TC 71-PSL. Prediction of service life of building materials and components. *Materials and Structures*, **22**, 385–92.

Matala, S. (1991) Service life model for frost resistance of concrete based on properties of fresh concrete, Nordisk Vägtekniska Förbundet, Broseminarium, Korsär, Denmark, 18 pp.

Parrot, L. (1992) Design for avoiding damage due to carbonation induced corrosion. April. Paper no. 62 – CEN TC 104/WG1/TG1/ Panel 1. Unpublished.

Pihlajavaara, S. E. (1984) The prediction of service life with the aid of multiple testing, reference materials, experience data, and value analysis. *VTT Symposium 48*, Espoo, Vol. 1, pp. 37–64.

Pihlajavaara, S. E. (1994) Contributions for the development of the estimation of long-term performance and service life of concrete. Helsinki University of Technology, Faculty of Civil Engineering and Surveying, Espoo, *Report 3*, 26 pp. + articles 49 pp.

RILEM (1986) Performance criteria for building materials. Final report of the RILEM committee 31-PCM. Technical Research Centre of

Finland, Espoo. *Research Notes 644*, 74 pp. + app. 36 pp.

Schiessl, P. (1976) Zur Frage der zulässigen Rissbreite und der erforderlichen Betondeckung in Stahlbetonbau unter besonderer Berücksichtigung der Karbonatisierung des Betons. Deutscher Ausschuss für Stahlbeton, Berlin, Vol. 255, 175 pp.

Sentler L. (1984) Stochastic characterization of carbonation of concrete. *Proceedings of the Third International Conference on the Durability of Building Materials and Components. VTT Symposium 50*, Espoo, pp. 569–80.

Siemes, A., Vrouwenvelder, A. and Beukel, A. van den (1985). Durability of buildings: a reliability analysis. *Heron*, **30**(3), 3–48.

Tuutti, K. (1982) Corrosion of steel in concrete. Swedish Cement and Concrete Research Institute, Stockholm. *CBI Research 4:82*, 304 pp.

Vesikari, E. (1981) Corrosion of reinforcing steels at cracks in concrete, Technical Research Centre of Finland, Espoo. *Research Reports 11/1981*, 39 pp. + app. 4 pp.

Vesikari, E. (1988) Service life of concrete structures with regard to corrosion of reinforcement. Technical Research Centre of Finland, Espoo. *Research Reports 553*, 53 pp.

Vesikari, E. (1994) Durability design of concrete structures with respect to frost attack. *Proceedings of the Fourth International Symposium on Cold Region Development.* 13–16 June. Association of Finnish Civil Engineers RIL, 2 pp.

Vesikari, E. (1995) Betonirakenteiden käyttöikämitoitus (Service life design of concrete structures). Association of Finnish Civil Engineers RIL, Helsinki. *RIL 183–4.9.* (In Finnish, 120 pp.)

Vuorinen, J. (1974) *Studies on Durability of Concrete in Hydraulic Structures.* Finnish Committee on Large Dams, Helsinki, 24 pp.

# RILEM

RILEM, The International Union of Testing and Research Laboratories for Materials and Structures, is an international, non-governmental technical association whose vocation is to contribute to progress in the construction sciences, techniques and industries, essentially by means of the communication it fosters between research and practice. RILEM activity therefore aims at developing the knowledge of properties of materials and performance of structures, at defining the means for their assessment in laboratory and service conditions and at unifying measurement and testing methods used with this objective. RILEM was founded in 1947, and has a membership of over 900 in some 80 countries. It forms an institutional framework for cooperation by experts to:

- optimize and harmonize test methods for measuring properties and performance of building and civil engineering materials and structures under laboratory and service environments;
- prepare technical recommendations for testing methods;
- prepare state-of-the-art reports to identify further research needs.

RILEM members include the leading building research and testing laboratories around the world, industrial research, manufacturing and contracting interests as well as a significant number of individual members, from industry and universities. RILEM's focus is on construction materials and their use in buildings and civil engineering structures, covering all phases of the building process from manufacture to use and recycling of materials.

RILEM meets these objectives though the work of its technical committees. Symposia, workshops and seminars are organised to facilitate the exchange of information and dissemination of knowledge. RILEM's primary output are technical recommendations. RILEM also publishes the journal *Materials and Structures* which provides a further avenue for reporting the work of its committees. Details are given below. Many other publications, in the form of reports, monographs, symposia and workshop proceedings, are produced.

Details of RILEM membership may be obtained from RILEM, École Normale Supéieure, Pavillon des Jardins, 61, avenue du Pdt Wilson, 94235 Cachan Cedex, France. RILEM Reports, Proceedings and other publications are listed below and details may be obtained from E & F N Spon, 2–6 Boundary Row, London SE1 8HN, UK. Tel: (0)171–865 0066, Fax: (0)171–522 9623, and on the Internet at http://www.thomson.com/chaphall/

*RILEM Reports*

1 **Soiling and Cleaning of Building Facades**
Report of Technical Committee 62-SCF. *Edited by L. G. W. Verhoef*
2 **Corrosion of Steel in Concrete**
Report of Technical Committee 60-CSC. *Edited by P. Schiessl*
3 **Fracture Mechanics of Concrete Structures - From Theory to Applications**
Report of Technical Committee 90-FMA. *Edited by L. Elfgren*
4 **Geomembranes - Identification and Performance Testing**
Report of Technical Committee 103-MGH. *Edited by A. Rollin and J. M. Rigo*
5 **Fracture Mechanics Test Methods for Concrete**
Report of Technical Committee 89-FMT. *Edited by S. P. Shah and A. Carpinteri*
6 **Recycling of Demolished Concrete and Masonry**
Report of Technical Committee 37-DRC. *Edited by T. C. Hansen*
7 **Fly Ash in Concrete - Properties and Performance**
Report of Technical Committee 67-FAB. *Edited by K. Wesche*

*RILEM Proceedings*

*RILEM Recommendations and Recommended Practice*

**RILEM Technical Recommendations for the Testing and Use of Construction Materials**

**Autoclaved Aerated Concrete - Properties, Testing and Design**
*Technical Committees 78-MCA and 51-ALC*

# Index

Milton Keynes UK
Ingram Content Group UK Ltd.
UKHW040053071024
449327UK00019B/532

9 780367 865375